Ageing and Life Extension of Offshore Structures

Ageing and Life Extension of Offshore Structures

The Challenge of Managing Structural Integrity

Gerhard Ersdal
University of Stavanger
Stavanger
Norway

John V. Sharp
Cranfield University
Cranfield
UK

Alexander Stacey
Health and Safety Executive
London
UK

The right of Gerhard Ersdal, John V. Sharp, and Alexander Stacey to be identified as the authors of this work has been asserted in accordance with law.

Registered Offices
John Wiley & Sons, Inc., 111 River Street, Hoboken, NJ 07030, USA
John Wiley & Sons Ltd, The Atrium, Southern Gate, Chichester, West Sussex, PO19 8SQ, UK

Editorial Office
The Atrium, Southern Gate, Chichester, West Sussex, PO19 8SQ, UK

For details of our global editorial offices, customer services, and more information about Wiley products visit us at www.wiley.com.

Wiley also publishes its books in a variety of electronic formats and by print-on-demand. Some content that appears in standard print versions of this book may not be available in other formats.

Library of Congress Cataloging-in-Publication Data

Names: Ersdal, Gerhard, 1966- author. | Sharp, John V., 1936- author. |
    Stacey, Alexander, 1959- author.
Title: Ageing and life extension of offshore structures : the challenge of
    managing structural integrity / Gerhard Ersdal, University of Stavanger,
    Stavanger, Norway, John V. Sharp, Cranfield University, Cranfield, UK,
    Alexander Stacey, Health and Safety Executive, London, UK.
Description: First edition. | Hoboken, NJ, USA : Wiley, [2019] | Includes
    bibliographical references and index. |
Identifiers: LCCN 2018042166 (print) | LCCN 2018042486 (ebook) | ISBN
    9781119284413 (Adobe PDF) | ISBN 9781119284406 (ePub) | ISBN 9781119284390
    (hardcover)
Subjects: LCSH: Offshore structures–Deterioration. | Offshore
    structures–Maintenance and repair. | Weathering of buildings.
Classification: LCC TC1670 (ebook) | LCC TC1670 .E77 2019 (print) | DDC
    627/.98–dc23
LC record available at https://lccn.loc.gov/2018042166

Cover Design: Wiley
Cover Image: © Wan Fahmy Redzuan/Shutterstock

Set in 10/12pt WarnockPro by SPi Global, Chennai, India

Printed in Singapore by C.O.S. Printers Pte Ltd

10 9 8 7 6 5 4 3 2 1

Contents

Preface

In the last decade life extension has been a dominant topic in the Norwegian and UK offshore industries, where all three authors have been involved.

This book is about the fundamental issues relevant to ageing of offshore structures and the necessary considerations for life extension. The aim of the book is to investigate and understand how these structures change with age and how these changes can be managed and mitigated.

The literature on structures is largely aimed at structural design despite the fact that particularly in UK and Norwegian waters over 50% of offshore structures are now in a life extension phase and are experiencing ageing. The literature on the management and assessment of these ageing structures is limited. This book is intended to help bridge that gap.

The opinions expressed in this book are those of the authors, and they should not be construed as reflecting the views of the organisations the authors represent. Further, the text in this book should not be viewed as recommended practice, but rather as an overview of important issues that are involved in the management of life extension.

The authors would particularly like to thank Narve Oma and John Wintle for carefully reviewing the manuscript, providing many valuable comments and making significant input to the content of this book. Further, the authors would like to thank Magnus Gabriel Ersdal and Janne N'jai for drafting some of the figures. The authors would also like to thank the helpful and patient staff at Wiley.

Harwell, Oxfordshire, April 2018

Gerhard Ersdal
John V. Sharp
Alexander Stacey

Definitions

The definitions given below apply to how they are used in this book.

Accidental limit state (ALS) A check of the collapse of the structure due to the same reasons as described for the ultimate limit state, but exposed to abnormal and accidental loading situations

Ageing A process in which the integrity (i.e. safety) of a structure or component changes with time or use

Air gap The positive difference between the lowest point of the underside of the lowest deck and the crest height of an extreme wave for a given return period (often 100 years)

As low as reasonably practicable (ALARP) This is a term often used in the regulation and management of safety critical systems.

Asset integrity management (AIM) This is the means of ensuring that the people, systems, processes and resources that deliver integrity are in place, in use and will perform when required over the whole life cycle of the asset

Barrier A measure intended to identify conditions that may lead to failure, hazardous and accidental situations, prevent an actual sequence of events occurring or developing, influence a sequence of events in a deliberate way, or limit damage and/or loss

Bilge The area on the outer surface of a ship's hull where the bottom shell plating meets the side shell plating

Design service life Assumed period for which a structure is to be used for its intended purpose with anticipated maintenance but without substantial repair from ageing processes being necessary

Duty holder A UK term for the operator in the case of a fixed installation (including fixed production and storage units); and the owner in the case of a mobile installation

Fatigue limit state (FLS) This is a check of the cumulative fatigue damage due to cyclic loads or the fatigue crack growth capacity of the structure

Fatigue Utilisation Index (FUI) This is the ratio between the effective operational time and the documented fatigue life

Fixed structure This is a structure that is bottom founded and transfers all actions on it to the sea floor

Flooded member detection (FMD) This is a technique which relies on the detection of water penetrating a member by using radiographic or ultrasonic methods

FPSO Floating production, storage, and offloading unit

FSO Floating storage and offloading unit

FSU Floating storage unit

Hazard Potential for human injury, damage to the environment, damage to property, or a combination of these

High Strength Steels (HSS) In this book defined as structural steels with yield strengths in excess of 500MPa

Hydrogen induced cracking (HIC) This is the process by which hydride-forming metals such as steel become brittle and fracture due to the introduction and subsequent diffusion of hydrogen into the metal

Jack-ups Mobile offshore units with a buoyant hull for transport and legs for supporting the hull onto the seabed

Life extension This is when the structure is used beyond its originally defined design life

Limit state This is a state beyond which the structure no longer fulfils the relevant design criteria

Management of change (MoC) This is a recognised process that is required when significant changes are made to an activity or process which can affect performance and risk

Microbiologically induced cracking (MIC) This is a form of degradation that can occur as a result of the metabolic activities of bacteria in the environment.

NDE Non-destructive examination

NDT Non-destructive testing

Partial safety factor For materials: this takes into account unfavourable deviation of strength from the characteristic value and any inaccuracies in determining the actual strength of the material. For loads: this takes into account the possible deviation of the actual loads from the characteristic value and inaccuracies in the load determination

Passive fire protection (PFP) These coatings are used on critical areas which could be affected by a jet fire. There are several different types which include cementitious and epoxy intumescent based

Performance standards Statement of the performance required of a structure, system, equipment, person or procedure and which is used as the basis for managing the hazard through the life cycle of the platform

Prestressing tendons High strength tendons are required to maintain the structural integrity of a concrete structure, particularly in the towers. These tendons are placed in steel ducts which are usually grouted following tensioning

Primary structure All main structural components that provide the structure's main strength and stiffness

Push-over analysis This is a non-linear analysis for jacket structures used for determining the collapse/ultimate capacity

Redundancy The ability of a structure to find alternative load paths following failure of one or more components, thus limiting the consequences of such failures

Reserve strength ratio (RSR) The ratio between the design loading (usually 100-year loading) and the collapse/ultimate capacity

Residual strength Ultimate strength of an offshore structure in a damaged condition

Robustness This reflects the ability of the structure to be damage tolerant and to sustain deviations from the assumptions for which the structure was originally designed

Safety critical elements (SCE) and Safety and environmental critical elements (SECE) These are those systems and components (e.g. hardware, software, procedures, etc.) that are designed to prevent, control, mitigate or respond to a major accident event that could lead to injury or death. This was further extended in the 2015 version of the UK safety case regulation to include environmental critical elements (SECE)

Scour Erosion of the seabed around a fixed structure produced by waves, currents, and ice

Secondary structure Structural components that, when removed, do not significantly alter the overall strength and stiffness of the global structure

Serviceability limit state (SLS) This is a check of functionalities related to normal use (such as deflections and vibrations) in structures and structural components

S–N curve This is a relationship between the applied stress range (*S*) and the number of cycles (*N*) to fatigue failure (regarding fatigue failure, see *Fatigue limit state*)

Splash zone Part of a structure close to sea level that is intermittently exposed to air and immersed in the sea

Stress concentration factor (SCF) Factor relating a nominal stress to the local structural stress at a detail

Structural integrity The state of the structure and conditions that influence its safety

Structural integrity management (SIM) This is a means of demonstrating that the people, systems, processes and resources that deliver structural integrity are in place, in use and will perform when required for the whole lifecycle of the structure with the aim of providing an acceptable safety level

Structural reliability analysis (SRA) This is used to analyse the probability of limit state failure of structures

Surveillance All activities performed to gather information required to assure the structural integrity, such as inspection of the condition and configuration, determining the loads, records, and document review (such as standards and regulations), etc.

Topsides Structures and equipment placed on a supporting structure (fixed or floating) to provide some or all of a platform's functions

Ultimate limit state (ULS) This is a check of failure of the structure of one or more of its members due to fracture, rupture, instability, excessive inelastic deformation, etc.

Water tight integrity The capability of preventing the passage of water through the structure at a given pressure head

Wave-in-deck Waves which impact the deck of a structure, which dramatically increase the wave loading on the structure

1

Introduction to Ageing of Structures

It is the destiny of the man-made environment to vanish, but we, short-lived men and women, look at our buildings so convinced they will stand forever that when some do collapse, we are surprised and concerned.

Levy and Salvadori (2002)

1.1 Structural Engineering and Ageing Structures

How long can a structure last? Historically we have seen structures failing before they were ready to be used.[1] Others, such as historical monuments, have lasted for centuries and millennia.[2] The life span of a structure will depend on its design, its construction, and fabrication, the material used, the maintenance performed, the challenging environment it has been exposed to, the accidental events it has experienced, and whether it is possible to repair and replace any damaged or deteriorated structural parts. Metallic structures from the 1700s are still carrying their intended loads. Such evidence may lead us to believe that structures may last forever. However, only one of the Seven Wonders of the Ancient World is still standing, namely the Great Pyramid of Giza (constructed around 2500 BCE).

Changes start to appear in structures from the moment they are constructed. The material in structures will degrade (mainly by corrosion and fatigue) and accumulate damage (such as dents and buckles). The environment the structures are placed in will change, and that will influence the degradation mechanism. The loads on a structure will change with changes in use. The foundations of the structure may experience settlement and subsidence, which implies additional stresses in the structures and may introduce changes to the loading. Furthermore, technological developments may lead to materials, equipment, and control systems related to the structure being outdated and spare parts for these systems becoming unavailable (obsolescence). Compatibility between new equipment and the equipment that is already in place on the structure (e.g. to control stability and ballast on a floating structure) may prove to be difficult. Ultimately we may face the problem of changing to a new technological solution with possible issues

1 Examples of this are the Cleddau Bridge (Milford Haven, UK) collapse during construction in 1970 and the *Wasa* ship that sank during launching in 1628.
2 Examples of this are the Caravan Bridge in Turkey (850 BCE), the Ponte Fabricio bridge in Rome (62 BCE), and the Pont du Gard aqueduct in France (18 BCE).

Ageing and Life Extension of Offshore Structures: The Challenge of Managing Structural Integrity,
First Edition. Gerhard Ersdal, John V. Sharp, and Alexander Stacey.
© 2019 John Wiley & Sons Ltd. Published 2019 by John Wiley & Sons Ltd.

concerning safety and functionality, or continue to use the old technological solutions with their limitations. All of the above may make a structure less safe.

The assessment of an ageing structure for possible further use has to be based on the available information. Ideally, information about the original design and fabrication of the structure, its use and the inspections performed over the years are required to determine whether a structure is fit for further use. This assessment needs to be based on an understanding of the current safety of the structure. However, for older structures, the necessary information required to show that they are sufficiently safe may be lost or impossible to obtain. Lack of information, new knowledge, and new requirements may change our understanding of the safety of a structure, and may force us to regard the structure as unsafe requiring further mitigation.

However, new knowledge, methods and requirements may provide information that leads to a better understanding of the integrity of an existing structure, including the possibility that the integrity is better than expected and sufficient for safe operation in the life extension phase.

Finally, as time passes since the design of a structure, the evolution of technical knowledge normally leads to society developing more stringent requirements for safety.[3] This improved understanding will increase expectations for the safe operation of structures, including older ones designed to lesser criteria.

Offshore structures are continuously exposed to all of the above types of change. They operate in an environment that causes corrosion, erosion, environmental and functional loads, incidents and accidents that deteriorate, degrade, dent, damage, tear, and deform the structure. In addition to the changes to the structures themselves, the loads and corrosive environments in which they operate will change over time. Further, the way these are used may change, which as a result will alter the loading, the environment these are exposed to and possibly the configuration of the installation. In addition, our knowledge about the structures will change, e.g. the type of information that we have retained from design and inspection of the structure. Further, the physical theories, mathematical modelling and engineering methods used to analyse the structures may change, typically as new phenomenon are discovered. Finally, our evaluation of offshore structures is also influenced by societal changes and technological developments. This may result in changes to the requirements that are set for offshore structures, taking into account obsolescence, lack of competence, and the availability of spare parts for old equipment.

These changes may be grouped into four different types:

- *Physical changes* to the structure and the system itself, their use, and the environment they are exposed to (condition, configuration, loading, and hazards).
- *Changes to structural information* (the gathering of more information from inspections and monitoring, but also potential loss of information from design, fabrication, installation, and use).

---

3 As an example, the number of traffic fatalities in Norway in the 1980s averaged 400 fatalities per year (0.01% of the population). Societal development has led to a lower acceptance of fatalities, and technological developments have led to safety improvements being possible, and the number of fatalities in 2015 reached a historic low of 125 (0.0025% of the population). At present, society expects further reduction in the number of traffic fatalities.

Figure 1.1 The four main elements of ageing of a structure.

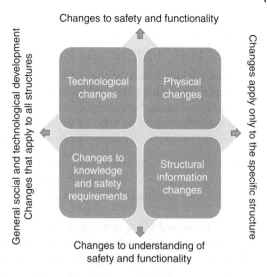

- *Changes to knowledge and safety requirements* that alter our understanding of the physics and methods used to analyse the structure, and the required safety that the structure is supposed to have.
- *Technological changes* that may lead to equipment and control systems used in the original structure being outdated, spare parts being unavailable, and compatibility between existing and new equipment and systems being difficult.

These groups of changes are illustrated in Figure 1.1, where it is indicated that the physical and technological changes impact the safety and functionality of the structure directly, while structural information changes and changes to knowledge and safety requirements primarily change how we understand the safety and functionality of a structure. Further, it is indicated that physical changes and structural information changes apply to one specific structure, while technological changes and changes to knowledge and safety requirements are a result of societal and technological developments, and are applicable to all structures.

These issues are highly relevant for a structural engineer; as we will show in Section 1.2, as the early offshore structures in the oil and gas industry are getting rather old. Many offshore structures from the 1990s are now passing their planned life expectance. However, there is a need for many of these structures to remain in service as there is still oil and gas remaining in the reservoirs. Further, many fixed and floating structures provide an important hub for the increasing number of subsea installations. The continued use of these older structures has the potential to save substantial costs and minimise environmental damage by avoiding the building of new structures.

In Section 1.3 we will show that failure statistics for structures indicate that structures in the oil and gas industry have a significant failure rate, particularly for floating structures. Further, older structures fail more often compared with newer structures. This is not surprising taking into account that structures will degrade and accumulate damage, that their use may change in unfavourable ways, that systems related to the structures may experience obsolescence and that newer structures may be designed according to improved methods and more stringent regulations and standards.

Facing the challenge of having relatively many older structures in the oil and gas industry, and at the same time knowing that older structures fail more often than newer ones, structural engineers need to:

- Understand how structures change as they get older (Chapter 3).
- Develop methods to assess these structures properly so that the structures that are unfit for further service are decommissioned, either because they are unsafe or they cannot be proved to be safe due to lack of important information (Chapter 4).
- Manage these older structures properly in their life extension phase (Chapter 5).

This book is generally about these items, but in order to understand older structures it is important to know about early designs and maintenance practices, as these will have an impact on our understanding of older structures. Similarly, it is important to know about the present requirements, because older structures will in many regions of the world be measured to the same safety standards as new structures. Further, the design of early structures was based on the knowledge and experience at that time and the methods often resulted in safe designs. In the intervening period there have been significant improvements in knowledge and experience which can be applied to the management of these older structures. These topics are covered in Chapter 2.

1.2 History of Offshore Structures Worldwide

Over the years, several types of offshore platforms have been used to produce oil and gas. One of the earliest successful fixed platforms was a wooden platform used by Pure Oil (now Chevron) and Superior Oil (now ExxonMobil) 1 mile from the coast in a water depth of 4.3 m in 1937 (Offshore 2004). The first floating production was from around the same time, using steel barges on which drilling rigs were installed. These barges were ballasted to rest on the bottom for drilling. When the wells were completed, the barges could be refloated and towed away to new well sites. Fixed structures were typically built around the wells for protection and to provide a platform where the wells could be maintained and serviced (Offshore 2004). However, the birth of offshore technology (Clauss et al. 1992) occurred in the mid-1940s when two steel platforms were erected in the Gulf of Mexico (GoM). One of these platforms was built 18 miles off the Louisiana coast in 5.5 m of water in 1946 by Magnolia Petroleum (now ExxonMobil) and the other platform was built in 1947 in 6.1 m of water, also 18 miles off the Louisiana coast, by Superior Oil.

In the UK and Norwegian sectors, the present day's North Sea oil and gas production commenced in 1965 when BP's Sea Gem drilling rig found gas in the West Sole field. Subsequently, further discoveries were made in the West Sole field and the Viking gas field in 1965 and the Leman Bank, Indefatigable, Balder and Hewett gas fields in 1966. This was followed by a series of significant finds, including:

- In 1969, Phillips Petroleum's discovery of the Ekofisk field in the Norwegian sector and Amoco's discovery of the Montrose field in the UK sector were announced. The Ekofisk field has been one of the major oil producing fields in the Norwegian sector.
- In 1970 BP discovered the Forties field 110 miles east of Aberdeen, with the first production in 1975; this was one of the largest producing oil fields in the UK sector, with five fixed steel platforms.

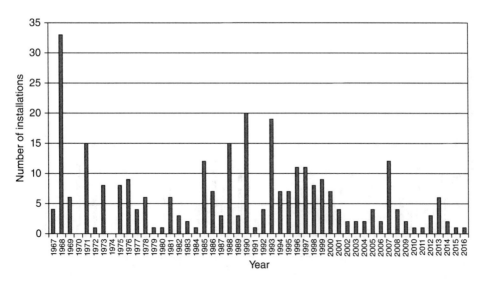

Figure 1.2 Fixed steel installations on the UK Continental Shelf by age.

- In 1971, Shell discovered the Brent field, located off the north east of Scotland. It has been one of the major oil and gas fields in the UK sector and some of the platforms are now being considered for decommissioning.

Further developments in the UK sector led to over 300 platforms, including 20 mobile installations, and 10 concrete platforms. The fixed steel platforms installed in UK waters are shown in Figure 1.2. Many of these are now operating beyond their original design lives.

The development of the Norwegian sector has led to over 100 platforms being installed, of which 15 were concrete. As noted for the UK sector, many of these installations are now in the life extension phase, see Figure 1.3.

In other parts of the North Sea, Denmark's first oil production began in 1972 from the Dan field. In Germany oil was found in the North Sea in 1981 in the Mittelplate field.

The types of platforms that are most used, and hence most relevant for life extension assessments, are shown in Appendix A and are:

- Fixed platforms:
  - Steel jacket structures (mostly pile supported or suction anchor supported)
  - Concrete gravity based structures
  - Jack-ups
- Bottom supported platforms:
  - Guyed towers, compliant towers, and articulated towers
- Floating platforms:
  - Semi-submersible platforms (mostly in steel, but one in concrete exists)
  - Tension leg platforms (mostly in steel, but one in concrete exists)
  - Ship-shaped platforms and barge shaped platforms (mostly in steel, but a few in concrete have been made)
  - Spar platforms

The most commonly used platform type is the steel jacket structure. These have been used in up to 400 m of water (Bullwinkle Jacket in the GoM), but are normally more

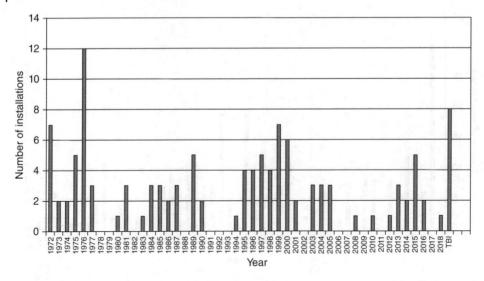

Figure 1.3 Age distribution of existing installations on the Norwegian Continental Shelf.

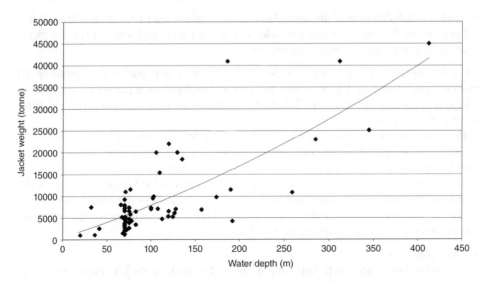

Figure 1.4 Jacket weight versus water depth for primarily jackets in northern European waters, but also including a few of the most known jackets in the GoM (based on publicly available data).

common in water depths up to 200 m as the weight of jackets in deeper water becomes very heavy (Figure 1.4). In terms of ageing and life extension (ALE), the majority of steel jacket structures are in water depths of less than 200 m.

Concrete gravity-based structures have been successfully used in up to 300 m of water. Since the 1970s, several gravity-based structures have been installed. The first were installed on the Ekofisk field in Norwegian waters in 1973 and the tallest was installed on the Troll field in Norwegian waters in 1995. A number of the earliest structures are now in the life extension phase.

A jack-up is a self-elevating unit using typically three legs to raise the hull above the surface of the sea on site, and with the ability to float using its buoyant hull from one location to another. The first jack-up was deployed in the mid-1950s. Jack-ups have played an important part in the discovery of oil and gas but also to a limited extent in the production of oil and gas. At present more than 500 jack-ups are in operation worldwide. Many of these jack-ups have been built with high strength steel in the legs, which may have special ageing issues, see Section 3.3.3.2.

Semisubmersible offshore platforms are the most widely used floating units. They are commonly used as mobile offshore drilling units, but are also used as crane vessels, offshore support vessels and as flotels. Around 50 semisubmersible platforms are used for production, and are permanently sited at one location. The earliest was the Argyll Floating Production Unit (FPU) in the UK North Sea in 1975. Many semisubmersible platforms are now in the ageing phase and especially the earlier designs do experience significant cracking.

Ship-shaped structures and barge shaped structures include floating storage units (FSUs), floating storage and offloading units (FSOs) and floating production, storage, and offloading units (FPSOs). Ship-shaped structures have the benefit of being able to store oil in the hull. Ships have been used for oil transport for many years, but the FSO Ifrika in the early 1970s was the first to be permanently used for production (Paik and Thayamballi 2007). Many of the Ship-shaped structures are now ageing and, like many ships, they experience various degradation mechanisms, see Section 1.3.5.

Today, the main areas of offshore oil and gas production are:

- the North Sea;
- the GoM;
- California (Ventura Basin – Santa Barbara Channel and Los Angeles Basin);
- Newfoundland and Nova Scotia;
- South America (Brazil [Campos and Santos Basins] and Venezuela);
- West Africa (Nigeria and Angola);
- Caspian Sea (Azerbaijan);
- Russia (Sakhalin);
- Persian Gulf;
- India (e.g. Mumbai High);
- South-east Asia;
- Western Australia.

The history of these offshore oil and gas producing regions extends to over 50 years and many structures are being used beyond their original design lives which, typically, range from 20 to 30 years. Improvements in the possible oil recovery from several fields have increased the interest for the continued use of these facilities well beyond their original design life. As noted earlier, another reason for extending the life of platforms is to provide a hub for the increasing use of subsea production 'satellites'.

The use of existing facilities beyond their original design life will in many cases be economically preferable, even if rather large modifications, refurbishments, repairs, and inspections have to be performed to achieve life extension. A major concern in this regard is that safety requirements should not be compromised.

In the North Sea, approximately two thirds of the infrastructure can be considered to be 'ageing' and in the life extension phase. The proportion of ageing structures is expected to increase, and there are plans to extend the life of some of these platforms by up to another 20 years. This trend is also reflected in many other regions, notably in the GoM, West Africa, the Middle East, etc., and consequently the management of ALE is an essential element of the integrity management process.

1.3 Failure Statistics for Ageing Offshore Structures

Whoever wishes to foresee the future must consult the past; for human events ever resemble those of preceding times.

Machiavelli

Those who cannot learn from history are doomed to repeat it.

Winston Churchill

We spend a great deal of time studying history, which, let's face it, is mostly the history of stupidity.

Stephen Hawking

1.3.1 Introduction

Offshore structures do fail. Particular examples of structural failure are *Sea Gem* that collapsed in 1965 (Adams 1967) due to fatigue and brittle failure of the suspension system, *Alexander L. Kielland* that capsized in 1980 (Moan et al. 1981) due to a fatigue failure and consequently the loss of 123 lives. Lessons learned from these failures are useful in order to avoid repeating the same failures again.

Regarding ageing structures, it is also specifically interesting to evaluate how the failure rate changes with age. Such data are not easily obtained, but some data are presented for offshore fixed steel structures in this section.

1.3.2 Failure Statistics of Offshore Structures

The International Association of Oil & Gas Producers (IOGP) has produced a document (IOGP 2010) on structural event statistics. Table 1.1 and Table 1.2 show statistics for all units and for floating units, respectively.

Total loss in this context means total loss of the unit from an insurance point of view. However, the unit may be repaired and put into operation again. Severe structural failure

Table 1.1 All units.

	Worldwide
Frequency of all severe structural failure (excl. towing)	4.55×10^{-5} per year
Frequency of severe structural failure caused by weather (excl. towing)	3.25×10^{-5} per year
Frequency of total loss (excl. towing)	4.55×10^{-5} per year
Frequency of total loss caused by weather (excl. towing)	1.30×10^{-5} per year

Table 1.2 Floating units (non-fixed units).

	Worldwide	UK
Frequency of all severe structural failure (excl. towing)	3.20×10^{-4} per year	1.09×10^{-2} per year
Frequency of severe structural failure caused by weather (excl. towing)	2.67×10^{-4} per year	3.28×10^{-3} per year
Frequency of total loss (excl. towing)	3.73×10^{-4} per year	
Frequency of total loss caused by weather (excl. towing)	1.07×10^{-4} per year	

events are when an installation loses its ability to support its topside as a result of operational and environmental loading.

Further, for fixed units the IOGP (2010) gives the frequency of all severe structural failures (excluding towing) which is estimated as 7.40×10^{-6} per year (worldwide estimate) and for UK the number is 1.09×10^{-3} per year.

No explanation for the rather large difference between worldwide data and UK data is provided. A possible reason may be the lack of reporting of accidents in some parts of the world.

1.3.3 Experience from Land Based Structures

Considering only land based steel structures, Oehme (1989) made a study considering the cause of damage and types of structures. This study is interesting because steel has been used in a wide variety of different structures. A total of 448 damage cases are reported. Approximately 98% occurred in the period of 1955–1984 and 62% in less than 30 years after installation. The cause of damage is presented in Table 1.3.

Table 1.3 Main causes of damage.

Damage cause (multiple denominations possible)	Totality		Buildings		Bridges		Conveyors	
	No.	%	No.	%	No.	%	No.	%
Static strength	161	29.7	102	33.6	19	14.8	40	36.0
Stability (local or global)	87	16.0	62	20.4	11	8.6	14	12.6
Fatigue	92	16.9	8	2.6	49	38.3	35	31.5
Rigid body movement	44	25	25	8.2	2	1.6	17	15.3
Elastic deformation	15	2.8	14	4.6	1	0.8	0	0
Brittle fracture	15	2.8	9	3.0	5	3.9	1	0.9
Environment[a]	101	18.6	59	19.4	41	32.0	1	0.9
Thermal loads	23	4.2	23	7.6	0	0	0	0
Others	5	0.9	2	0.7	0	0	3	2.7
Sum	543	100	304	100	128	100	111	100

a) For environment in this study corrosion is the primary meaning, but the term seems to also include other environmental degradation processes.

Source: Based on Oehme (1989).

Table 1.4 Causes of failure of structural components of bridges.

Cause	No.
Overload	8
Fatigue	12
Deterioration other than fatigue (corrosion, concrete cracking, etc.)	18

Source: Carper (1998).

As we are interested here in ageing structures, the mentioned degradation mechanisms of fatigue and environment (corrosion) are the first group of causes for failures that should receive attention. However, introduced actions due to e.g. differential settlements could play a part in several of the other causes of failure, but it is not possible to draw conclusions based on the available data.

Carper (1998) has studied causes of failure of structural components of bridges in Canada in the years 1987–1996. The causes found are shown in Table 1.4.

In all these 'failures', only overloads resulted in complete failure of the component. Fatigue failures were rarely complete failures because redistribution between remaining elements occurred as the crack grew. This shows the importance of redundancy in the evaluation of safety when fatigue occurs.

1.3.4   Experience from Offshore Fixed Steel Structures

Most structural failures of offshore oil and gas platforms have occurred in the GoM during hurricanes. Many of these platforms are significantly older than those on the Norwegian and UK continental shelves. An overview of the hurricanes in the GoM and the number of platforms destroyed is given in Table 1.5.

Data on the hurricanes prior to hurricane Andrew are difficult to find, and the majority of the failures have occurred from hurricane Andrew and subsequently. Hence, the data used in the following evaluations are from the following hurricanes:

- Andrew (PMB Engineering 1993)
- Lili (Puskar and Ku 2004)
- Ivan (Energo 2006)
- Kartina and Rita (Energo 2007)
- Gustav and Ike (Energo 2010)

Summing up the findings from these reports, the distribution shown in Figure 1.5 is illustrated for platform vintage at damage or collapse.

To gain information of the age distribution of platforms in the GoM, the report 'Forecasting the number of offshore platforms in the Gulf of Mexico OCS to the year 2023' by Pulsipher et al. (2001) has been used, as shown in Figure 1.6.

Taking into account the number of platforms of different age, corrected for the date shown, a relative probability of failure may be determined, as shown in Figure 1.7. The relative probability is calculated by dividing the number of structural failures for a group

Table 1.5 Historic damage to offshore fixed platforms from hurricanes.

Hurricane	Year	Platforms destroyed
Grand island[a]	1948	2
Carla[a]	1961	3
Hilda[a]	1964	14
Betsy[a]	1965	8
Camille[a]	1969	3
Carmen[a]	1974	2
Frederic[a]	1979	3
Juan[a]	1985	3
Andrew[a]	1992	28
Lili[a]	2002	7
Ivan[a]	2004	7
Katrina[a]	2005	46
Rita[a]	2005	68
Gustav[b]	2008	1
Ike[b]	2008	59

a) Data from Energo (2006).
b) Data from Energo (2010).

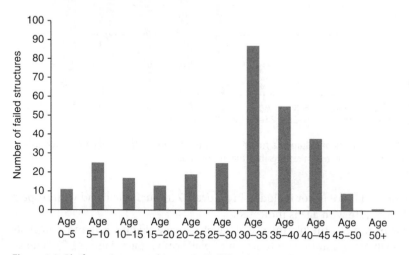

Figure 1.5 Platform vintage at damage in GoM hurricanes.

of platforms with a certain vintage (Figure 1.5) and dividing them by the number of reported structures of this age (Figure 1.6).

Further studies need to be performed to determine if the definition of major structures is the same in all of these reports. However, it is assumed that the age distribution curve is representative, but the actual numbers need to be studied further. Hence, the

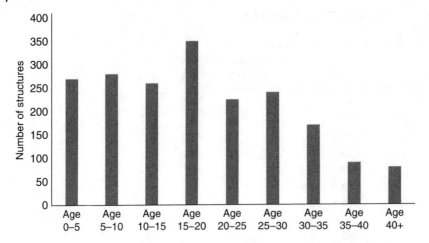

Figure 1.6 Age distribution of fixed major structures in 1997. Source: Based on Pulsipher et al. (2001).

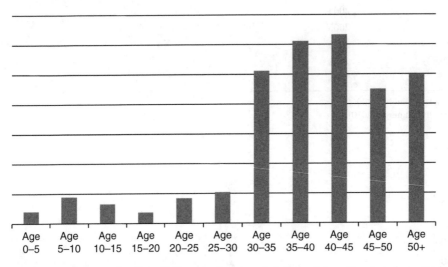

Figure 1.7 Trend in probability of failure as a function of platform vintage (numbers for vertical axis are purposely not included due to limitations in actual numbers).

actual probability of failure is not included in Figure 1.7, though the trend with respect to increased failure rate with age is indicated.

It is apparent from both the platform vintage at damage (Figure 1.5) and especially the relative probability of failure as a function of platform vintage (Figure 1.7) that age is an important factor, with a significant change in the relative probability of failure for platforms older than 30 years.

However, the reports on these damages and collapses in the hurricanes described in Table 1.5 do not clearly imply physical ageing (deterioration) as an important factor for most of these failures. As noted later, inadequate design may have been the primary cause. In fact, very little prior degradation and inspection findings are included in the reports on these accidents. One example is mentioned of a platform exposed to a boat collision prior to it collapsing in a hurricane, possibly indicating that the damages from this boat collision could have contributed to the collapse in the hurricane. In other cases, corrosion of primary steelwork was also known and contributed to a reduction of resistance. More studies which are more specifically on this issue may reveal more platforms that were degraded prior to their collapse in these hurricanes.

An important cause for several of these accidents may be that the use of older standards for design and assessment has been accepted for older structures. This is further discussed by Energo (2007) which analyses platforms according to the relevant American Petroleum Institute (API) standard that was used in their design. Platforms that were designed according to standards earlier than the 9th edition of API-RP2A showed a higher failure rate.

Most damage to platforms in the hurricanes resulted in leaning (between 1° and 45°) as shown in Figure 1.8, or toppling, as a result of member or joint failure and in most cases multiple failures.

It should be mentioned that a significant cause of the hurricane losses has been that the wave heights experienced were far greater than that had been anticipated at the design stage but other ageing mechanisms could also have contributed. It should also be noted that the majority of failures presented in Figure 1.7 are partly from hurricanes occurring after 1997, while the age distribution of platforms in the GoM as presented in Figure 1.6 is from 1997. Hence, solid conclusions should not be drawn e.g. on frequency of failure, but the indication of a trend is interesting and invites further investigation. Further, the results obtained by using the failure statistics from Figure 1.7 and normalising them by the numbers of platforms in each age group as shown in Figure 1.6 assumes that the platforms in all age groups have been equally affected by the hurricanes. This may not be the case but there are no data indicating that the exposed platforms belong to a

Figure 1.8 Gulf of Mexico platform damaged in a hurricane. Source: Gerhard Ersdal.

particular age distribution. Hence, again the findings presented should be viewed as an indication of a trend and not as background for actual frequency of failure for each age range.

1.3.5 Experience from the Shipping and Mobile Offshore Unit Industries

Studies into the effect of ageing have also been performed in the shipping and MOU (mobile offshore unit) industries. Similar trends as seen for fixed structures in the GoM are also found here. Figure 1.9 summarises the types of damage to critical hull members. The damage includes corrosion, cracking of the structure, vibration, and other types (e.g. collision causing buckling). Damage is defined here as defects or deterioration requiring repairs. Figure 1.9 gives the trend in the damage with age.

As stated in SSC (2000), it can be seen that damage due to corrosion accounts for more than half the total reported and that corrosion increases with the age of a ship. The levelling out at later life may be influenced by the fact that there are fewer ships of that age. Figure 1.10 summarises the corrosion and fatigue related damage in cargo tanks, ballast tanks, and other spaces in ships as a function of ship age.

Similarly, studies have been undertaken on MOUs operating primarily in Norwegian waters, specifically the Aker H-3 type MOUs. Figure 1.11 indicates the average number of cracks on the studied MOUs, defined as the occurrence of a crack detected in the five-year in-service inspection programme.

Figure 1.11 shows a clear trend of increasing number of cracks with age for MOUs. However, the trend may not be as easily detectable if one looks at the ageing of a particular MOU, primarily due to the limited data available. Statoil (2002) should be consulted for detailed discussion on these cases, but a clear trend towards higher incident rates is found for older facilities. For both ships and MOUs it seems to be primarily physical ageing that is studied, and hence this is considered to be the cause of the incidents.

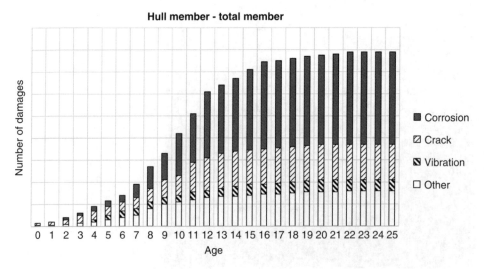

Figure 1.9 Damage to hull structural members by different causes and ship age for all ship types. Source: Based on SSC (1992, 2000).

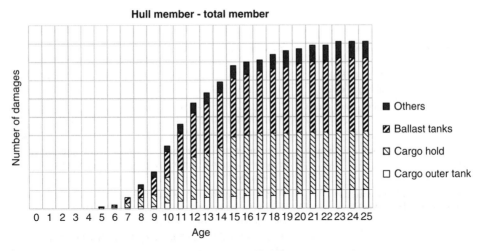

Figure 1.10 Damage due to corrosion and fatigue for different locations and ship age. Source: Based on SSC (1992, 2000).

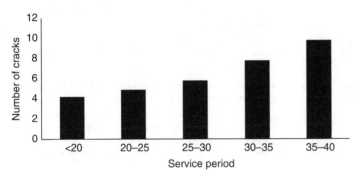

Figure 1.11 Average number of cracks (five-year inspection intervals) on typical Aker H-3 MOUs. Source: Based on Statoil (2002).

1.4 The Terms 'Design Life' and 'Life Extension' and the Bathtub Curve

> *A building is conceived when designed, born when built, alive while standing, dead from old age or an unexpected accident. The accidental death of a building is always due to the failure of its skeleton, the structure.*
>
> Levy and Salvadori (2002)

The term 'design life' of an offshore facility (or a land based building) is not well defined in codes and standards. Table 1.6 shows several different interpretations of this.

For our purposes, the definitions given in ISO 2394 (ISO 1998) and ISO 19902 (ISO 2007) are probably the most relevant.

However, we propose a slightly modified form which may be used for practical purposes for ageing: 'The design life is the assumed period for which a structure is to be used

Table 1.6 Definitions of design life in codes and standards.

Code, standard, or guidance	Definition of design life
ISO 19900 (ISO 2013) – Offshore structures – General requirements	Section 3.5 – Service requirements – the expected service life shall be specified in design
ISO 19902 – Fixed steel installations	Section 4 – The assumed period for which a structure is to be used for its intended purpose with anticipated maintenance but without substantial repair being necessary
NORSOK – N-001 (Standard Norge 2012)	Structures shall be designed to withstand the presupposed repetitive (fatigue) actions during the life span of the structure
HSE Design and Construction Regulations (HSE 1996)	Reg. 4 – Need to ensure integrity of a structure during its life cycle. Processes of degradation and corrosion to be accounted for at the design stage
	Reg. 8 – Need to maintain integrity of structure during its life cycle
ISO 2394 – General principles on reliability for structures	Definition of design working life: 'The assumed period for which a structure is to be used for its intended purpose with anticipated maintenance but without substantial repair being necessary'
DNV – Classification Note 30.6, Structural reliability methods	Definition of design life: 'The time period from commencement of construction to until condemnation of the structure'
Department of Energy/HSE Guidance Notes (HSE 1994)	Calculated fatigue life should not be less than 20 years, or the required service life if this exceeds 20 years
	The (cathodic protection) current to all parts of the structure should be adequate for protection for the duration of the design life

for its intended purpose with anticipated maintenance but without substantial repair from ageing processes being necessary'.

The original design life is the design working life of the installation assumed at the time of design. Changes in use and/or structural modification, if they occur within the original design life, are considered as 'reassessment and requalification'. A reassessment may lead to a revision of the (remaining) 'design working life', but is generally based on the initial design assumptions and criteria.

To illustrate this, a refined version of the bathtub curve is shown in Figure 1.12, with an initial phase, the maturity phase representing the useful life and the ageing and terminal phases representing the first and second part of the end of life (HSE 2006).

The (original) design life is assumed to be the period during which the structure can safely be used, and hence must be assumed to reach somewhat into the maturity phase but not into the ageing phase. Life extension is when the structure is used beyond this originally defined design life. A complication to this can arise if there is a difference between the:

- design life as defined in design specifications (by the owner/duty holder);
- original calculated design life (by the designer); and

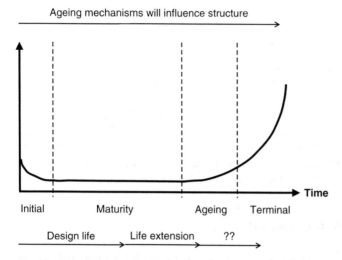

Figure 1.12 Ideally, a representation of design life and life extension related to the bathtub curve. However, it should be noted that ageing may set in earlier if the structure is not managed properly. Source: Based on HSE (2006).

- updated calculations of life (with the most up to date methods and computers at a later stage).

Here, life extension is defined as the stage from the original design life as defined in the design specification.

The phases as indicated in Figure 1.12 may be understood as (HSE 2006):

- Stage 1: 'Initial'
  As structures, systems and equipment are put into service there may be a relatively higher rate of damage accumulation and issues requiring attention as a result of inherent weakness or faults in the design, materials, or fabrication – and bedding-in effects.
- Stage 2: 'Maturity'
  After the structures, systems and equipment have passed through the early-life problems, they enter the second stage. This longer 'maturity' stage is when the equipment is predictable, reliable and is assumed to have a low and relatively stable rate of damage accumulation and few issues requiring attention. It is operating comfortably within its design limits.
- Stage 3: 'Ageing'
  By this stage the structures, systems, and equipment have accumulated some damage and the rate of degradation is increasing. Signs of damage and other indicators of ageing are starting to appear. Further, it becomes more important to determine quantitatively the extent and rate of damage and to make an estimate of remnant life. Design margins may be eroded and the emphasis shifts towards fitness-for-service and remnant life assessment of specific damaged areas.
- Stage 4: 'Terminal'
  As accumulated damage to structures, systems and equipment becomes increasingly severe, it becomes clear that the structures, systems and equipment will ultimately need to be repaired, refurbished, decommissioned, or replaced. The rate

of degradation is increasing rapidly and is not easy to predict. In this final 'terminal' stage of the equipment's life, the main emphasis is on guaranteeing adequate safety between inspections while keeping the equipment in service as long as possible.

The concept of a design life is easiest to interpret for a fixed structure remaining on one site during its life. For a mobile installation that may operate in different parts of the world subjected to different environmental conditions, the interpretation of design life and assessment of ageing becomes more difficult. Some classification societies have developed methods for assessing and managing, for example fatigue life for such structures (such as the fatigue utilisation index in DNVGL), see Section 4.5.3. Alternatively, dedicated fatigue analysis from the different locations of operation may be carried out consecutively in order to obtain the status on accumulated fatigue life.

1.5 Life Extension Assessment Process

The main objective of a life extension assessment is to ensure that ageing facilities are still sufficiently safe. To achieve this objective, the process shown in Figure 1.13 may be used (Ersdal et al. 2008). Additional significant work relevant to the assessment for life extension can also be found in Sharp et al. (2011).

Definition of context and objective for life extension. As a basis for the structural assessment, a key input is the owner's expectations for how long the structure is planned to be

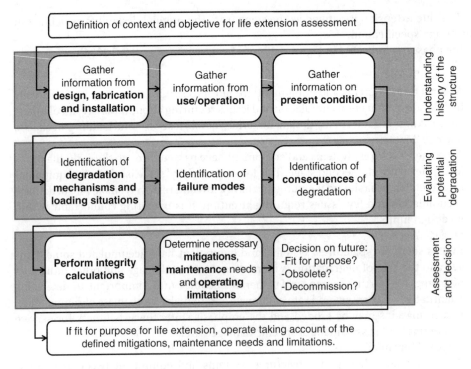

Figure 1.13 Life extension assessment process.

used. The context should also include the regulations and standards that should be used for the assessment of the ageing structure for life extension. Another important input is the way the structure will be used, including any planned changes. Such information is the preliminary requirement for the following processes.

Understanding of the history of the structure. This involves the evaluation of the degradation history of the structure and its components and determining the present condition. The assessment of the structure should be based on the as-is condition, updated drawings, computer models, and analysis of the as-is condition to ensure that all changes to the as-built condition and any new technology are incorporated in the new analysis. This also includes the impact of any damage to and degradation of the structure.

This stage should include gathering the necessary as-is information on the structure and marine systems (e.g. correct drawings, thickness measurements, etc.).

Statistics on actual incidents and accidents will influence the risk analysis which provides the basis of design accident specifications for the structure. Any successful performance history (such as the absence of cracks during inspections) will also be important documentation for reducing uncertainty about the structure. With limited information on performance history, which often applies to older structures, uncertainty will be a major challenge and life extension will ideally need to be based on significantly higher safety factors.

Evaluation of the potential degradation and loading. It is necessary to evaluate which aspects of ageing that may reduce safety. This will include identifying possible ageing related degradation mechanisms and failure modes which may affect the structure. An important aspect is whether the past or future rate of degradation has been or is expected to be slow or rapid. The history of incidents and accidents on the unit should also be evaluated, along with their influence on such aspects as the strength of the structure. Another important aspect is to identify if degradation can be found by inspection, if present, and if the structural part can be repaired.

Assessment. The structure's integrity should be evaluated taking into account its as-is condition, its assumed future performance, its use, and proposed modifications and mitigations. Typically, this will include checks of structural strength, fatigue life, and whether corrosion protection is sufficient.

A few standards give some guidance on the assessment of structures and marine systems for life extension. One example of such a standard is the NORSOK N-006 standard (Standard Norge 2015). Assessment for life extension is discussed in more detail in Chapter 4 and mitigations are outlined in Chapter 5.

Decision on further use. Ultimately, the question must be whether to continue using the structure in life extension or to decommission it. The answer depends on whether a combination of mitigating actions and modifications will be sufficient to demonstrate compliance with national safety requirements and whether an economic case can be made for life extension.

The economic considerations are illustrated in Figure 1.14 showing how the balance between revenue and costs changes with time and the uncertainty associated with this. This leads to a period of time which could constitute 'end of life' from economical considerations, unless other factors become important, such as decommissioning costs and the opportunity for re-use or tie-ins.

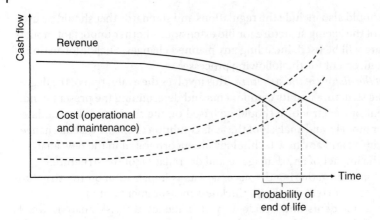

Figure 1.14 Balance between revenue and cost with time, indicating timescale associated with economic end of life. Source: Based on Stacey (2010).

Integrity management of structure in life extension phase. A life extension assessment needs to result in an updated plan for structural integrity management (see Section 2.4), taking into account the ageing effects that the structure is assumed to be exposed to.

Bibliographic Notes

Section 1.1 is based on Ersdal (2014). Section 1.2 is based on general information in the public domain, the authors' personal experience and notes. Section 1.3 is partly based on Ersdal et al. (2014). Section 1.4 is based on HSE (2006) and Ersdal (2014). Section 1.5 is based on Ersdal et al. (2008).

References

Adams, J.R. (1967). Inquiry into the Causes of the Accident to the Drilling Rig Sea Gem. The Ministry of Power, HMSO CM3409, London.

Carper, K. (1998). Conference on failures in architecture and engineering: What are the lessons? EPFL, Lausanne.

Clauss, G., Lehmann, E., and Østergaard, C. (1992). *Offshore Structures, Vol. 1 Conceptual Design and Hydromechanics*. Springer Verlag.

Energo. (2006). Assessment of fixed offshore platform performance in Hurricanes Andrew, Lili and Ivan. MMS Project no. 549.

Energo. (2007). Assessment of fixed offshore platform performance in Hurricanes Kartina and Rita. MMS project no. 578.

Energo. (2010). Assessment of fixed offshore platform performance in Hurricanes Gustav and Ike. MMS project no. 642.

Ersdal, G. (2014). Ageing and life extension of structures, Compendium at the University of Stavanger.

Ersdal, G., Kvitrud, A., Jones, W., and Birkinshaw, M. (2008). Life extension for mobile offshore units require robust management: how old is too old? *Journal of International Association of Drilling Contractors* 64 (5): 54–58.

Ersdal, G., Sharp, J., and Galbraith, D (2014). Ageing accidents – suggestion for a definition and examples from damaged platforms. OMAE14, San Francisco, CA.

HSE (1994). *Guidance Notes for the Design and Construction of Offshore Structures*, 4e. Health and Safety Executive (HSE).

HSE. (1996). *The Offshore Installations and Wells (Design and Construction, etc) Regulations*. Health and Safety Executive (HSE) SI 1996/913.

HSE. (2006). Plant ageing – Management of equipment containing hazardous fluids or pressure. HSE RR 509.

IOGP. (2010). International Association of Oil & Gas Producers (IOGP) Risk Assessment Data Directory. Report No. 434. IOGP, London.

ISO. (1998). ISO 2394. General principles on the reliability for structures. International Standards Organisation.

ISO. (2007). ISO 19902. Petroleum and natural gas inductries – Fixed steel offshore structures. International Standards Organisation.

ISO. (2013). ISO 19900. Petroleum and natural gas industries – General requirements for offshore structures. International Standards Organisation.

Levy, M. and Salvadori, M. (2002). *Why Buildings Fall Down*. New York: W.W. Norton & Company.

Moan, T., Bekkvik, P., and Næsheim, T. (1981). 'Alexander L. Kielland'-ulykken. Norges Offentlige Utredninger, NOU 1981:11, Universitetsforlaget, Oslo, Norway. (In Norwegian with an English summary.)

Oehme, P. (1989). Schäden an Stahltragwerken – eine Analyse (Damage analysis of steel structures). *IABSE Proceedings P-139/89*.

Offshore. (2004). Special Anniversary – the History of Offshore: Developing the E&P Infrastructure, https://www.offshore-mag.com/articles/print/volume-64/issue-1/news/special-report/special-anniversary-the-history-of-offshore-developing-the-ep-infrastructure.html (accessed 1 April 2018).

Paik, J.K. and Thayamballi, A.K. (2007). *Ship-Shaped Offshore Installations: Design, Building and Operation*. Cambridge University Press.

PMB Engineering (1993). *Hurricane Andrew – Effects on Offshore Platforms – Joint Industry Project*. San Francisco, CA.: PMB Engineering, Inc.

Pulsipher, A.G., Omowumi, O.I., Mesyanzhinov, D.V. et al. (2001). *Forecasting the Number of Offshore Platforms on the Gulf of Mexico OCS to the Year 2023*. Prepared under MMS Contract 14-35-0001-30660-19934 by Center for Energy Studies, Louisiana State University, Baton Rouge, LA. Gulf of Mexico OCS region: U.S. Department of the Interior, Minerals Management Service.

Puskar, F.J. and Ku, A.P. (2004). Hurricane Lili's impact on fixed platforms and calibration of platform performance to API RP 2A. OTC paper 16802.

Sharp, J.V., Wintle, J.B., Johnston, C., and Stacey, A. (2011). Industry Practices for the Management of Ageing Assets Relevant to Offshore Installations. Paper no. OMAE2011–49264

SSC. (1992). Marine Structural Integrity Programs (MSIP). Ship Structure Committee report no. 365.

SSC. (2000). Prediction of Structural Response in Grounding Application to Structural Design. Ship Structure Committee report no. 416.

Stacey, A. (2010). HSE Ageing and life extension inspection programme. Presentation at the Norwegian Petroleum Safety Authority Workshop on Ageing and Life Extension, PSA, Norway (7 April 2010).

Standard Norge. (2012). NORSOK N-001 Integrity of offshore structures. Edition 8. Standards Norway, Lysaker, Norway.

Standard Norge. (2015). NORSOK N-006 Assessment of structural integrity for existing offshore load-bearing structures. Edition 2. Standards Norway, Lysaker, Norway.

Statoil. (2002). Ageing and operability project – Mobile drilling units. Document no. 02.2001.BRT RIG. Revision date 8 January 2002.

2

Historic and Present Principles for Design, Assessment and Maintenance of Offshore Structures

2.1 Historic Development of Codes and Recommended Practices

In order to understand an old structure, it is important to understand the standards, calculation methods, materials, fabrication methods, installation methods and maintenance routine that was used at the time when the structure was built.

2.1.1 US Recommended Practices and Codes

The American Petroleum Institute (API) issued in 1969 its first recommended practice (RP 2A) for the design of offshore structures, with the first set of design procedures for the offshore industry (API 1969). In this first edition of RP 2A, a limited 25-year return wave criterion was recommended, which has since been increased. In 1972 (fourth edition), upgraded procedures for the design of tubular joints were provided (API 1972). In 1977 (ninth edition), the reference wave was introduced and the return period was increased from the 25-year to a 100-year return period and the tubular design procedures were upgraded (API 1977). In 1982 (13th edition), further improvements to tubular joint design provisions were introduced (API 1982). In 1980, API started developing the new Load Resistance Factored Design (LRFD) method for fixed offshore platform design which was eventually published in 1993 (API 1993). This 20th edition of API RP 2A also presented a completely revised wave force formulation as the basis for design.

The API tubular joint static design technology has been under continuous development since the first edition of API RP 2A in 1969. In the fourth edition, issued in 1972, some simple recommendations were introduced based on punching shear principles. In the fourth edition, factors were introduced to allow for the presence of load in the chord and the brace-to-chord diameter ratio. In the ninth edition, issued in 1977, differentiation was introduced in the allowable stress formulations for the joint and loading configurations, i.e. between T/Y, X, and K joints. In the 14th edition, the punching shear stress formulations were considerably modified and included a more realistic expression to account for the effect of chord loads as well as providing an interaction equation for the combined effect of brace axial and bending stresses. The static strength guidance then essentially remained unchanged for all editions up to the 21st.

Ageing and Life Extension of Offshore Structures: The Challenge of Managing Structural Integrity, First Edition. Gerhard Ersdal, John V. Sharp, and Alexander Stacey.
© 2019 John Wiley & Sons Ltd. Published 2019 by John Wiley & Sons Ltd.

The tubular joint fatigue design procedures have also been upgraded. The welded joint X and X′ fatigue design curves in the RP 2A editions from nos. 11–21 have since been replaced by a basic $S–N$ curve with a slope $m = 3$ that changes to 5 at 10 million cycles. Fatigue life correction factors for several factors, such as seawater, thickness and use of weld profile control, grinding and peening, were also introduced. Improved equations to evaluate stress concentration factors (SCFs) at tubular joints were also established.

Most installations in the US now in the life extension stage were designed to earlier versions of API RP 2A with the limitations described above.

The 22nd edition of API RP 2A issued in 2014 (API 2014a) saw the introduction of a supplement named API RP 2SIM (API 2014b) aimed at the ongoing structural integrity management (SIM) which is increasingly important. This new section, however, did not address life extension. This edition also included substantial revisions to the provisions related to strength and fatigue performance of tubular joints.

In addition, a new document named API RP 2FB was published in 2014 to provide new guidelines on the analysis and design of offshore platform topsides against fire and blast.

Initially in other countries which were developing offshore structures, there was a lack of appropriate design standards and as a result API RP 2A became the default used as the basis for design of such structures around the world. By the end of the 1970s, other standards and codes were introduced. The most notable of these was the issue of the UK Department of Energy Guidance Notes (see below) and the Norwegian Petroleum Directorate (NPD) rules supported by NORSOK standards, see e.g. Standard Norge (2012). In addition, several ship classification societies became involved in the offshore industry and released appropriate rules, some based on API RP 2A. These included the American Bureau of Shipping, Det Norske Veritas, and Lloyds Register.

2.1.2 UK Department of Energy and HSE Guidance Notes

Over the years the development of standards for UK waters has been heavily influenced by several accidents; the first of these was the *Sea Gem* accident in December 1965 when the rig was exploring for gas in the Southern North Sea (Adams 1967). The rig collapsed with 32 men on board, rapidly sinking and with 13 fatalities. This alerted the Government to the harsh conditions of working in the North Sea and the need for more robust safety legislation. As a result, the Minerals Workings (Offshore Installations) Act was introduced in 1971 by the UK Department of Energy. Following this, the Offshore Installations (Construction and Survey) Regulations were implemented in 1974. As required by these regulations, the Certification regime was introduced, entailing regular certification by an appointed organisation (e.g. Lloyds Register, Det Norske Veritas). These were supported by a set of Guidance Notes issued by the Department of Energy for the regulation of the offshore industry. Most UK installations now in the life extension stage were designed and initially operated under the Certification regime.

The first set of Guidance Notes entitled 'Offshore Installations: Guidance on Design, Construction and Certification' was fairly basic which included chapters on steel materials and primary structure. The second issue in 1977 (Department of Energy 1977) extended the guidance and this was further developed in the third edition in 1984 (Department of Energy 1984). The fourth and last edition was published in 1990

(HSE 1990). Early fatigue design for welded tubular joints was much less conservative than current guidance; it was derived partly from fatigue data for welded plates tested in air. The 1977 version of the Guidance Notes was based on the 'Q' $S-N$ curve associated with a hot spot stress and limited safety factors. In the early 1980s, the Department of Energy set up a panel to draft new fatigue guidance which led to the introduction of the T curve for tubular joints in 1984 as part of the third edition of the Guidance Notes (Department of Energy 1984). This curve was based on test data for joints from several large programmes which included tests on 300 welded plates and 50 tubular joints as part of the UKOSRP-II programme (Department of Energy 1987). A correction factor for joint thickness was also included, based on the fatigue performance of welded plates of different thicknesses. The effect of environment on performance (factor of 2 reduction with cathodic protection) was also based on the results from the testing of welded plates in seawater both with and without cathodic protection. Since existing $S-N$ curves for tubular joints are based on hot spot stresses, the calculation of the SCFs is accepted as an important requirement for the prediction of fatigue lives. In the 1977 Guidance Notes, no recommendations were provided on the suitability of equations that had been developed to predict SCFs. As part of the UKOSRP-II programme, assessment of existing parametric equations showed that several could lead to serious underpredictions of fatigue life. As a result, recommendations were made on preferred equations.

Structural design criteria improved with each set of Guidance Notes and the fourth edition contained detailed static strength formulations and further improvements in fatigue requirements for tubular joints. The fatigue section in the fourth edition was developed to be the most comprehensive requirement available at that time with the introduction of a new $S-N$ curve for tubular joints (T') based on the latest available test data and became the basis for the International Standards Organisation (ISO) standard (see below).

A very serious accident occurred in UK waters in July 1988 when the *Piper Alpha* platform suffered a major fire and explosion, killing 165 workers, the worst accident in the offshore industry. This led to the Cullen inquiry which produced over 100 recommendations for the safer management of the UK offshore industry. The previous regulations were considered to be too prescriptive and one of the recommendations was that the Construction and Survey Regulations should be revoked and replaced with a series of new regulations based on the management of risk through a safety case covering key issues such as fire and explosion and evacuation, escape and rescue, which were the key failure issues in the *Piper Alpha* disaster. The safety case regime was introduced in 1992 and fully implemented in 1995.

Subsequently, the Guidance Notes were withdrawn as the new regime placed responsibility on the operator for the safe management of offshore installations and the codes and standards that were appropriate for this. As a result, the UK offshore industry now uses mainly ISO standards, supplemented by other standards when required.

More recently, the very serious oil spill and accident in the Macondo field in the Gulf of Mexico and death of 11 workers in 2010 has led the European Union (EU) to be concerned that all EU countries did not meet high enough safety and environmental offshore standards. As a result, the EU produced a Directive (EU 2013) aimed at a safer and more environmentally friendly operation in European offshore waters. In response to this, the UK Government has modified its safety case regime to introduce a requirement to introduce 'environmental critical elements' such that they provide protection against an environmental event.

2.1.3 Norwegian Standards

As already mentioned, the API issued in 1969 its first recommended practice (RP 2A) for the design of fixed offshore structures, with the first set of design procedures for the offshore industry. Further revisions followed as described above. The first structures in Norway were designed according to these early API standards.

The NPD was the regulator in Norway from 1972. The first regulation for the 'Structural design of fixed offshore structures' was published in 1977 (with a proposal for this regulation published in February 1976). This is generally regarded as the first partial factor based limit state design code for offshore structures. In the 1977 version of the NPD regulation (NPD 1977), limit states for serviceability, ultimate strength, fatigue, and progressive collapse were included. Environmental loads for ultimate limit state (ULS) were specified with a return period of 100 years. In addition, accidental loads were included but with no specified probability of the accidental loads (characteristic values to be subject to approval by NPD). In 1977, DNV also published their codes for 'Design construction and inspection of offshore structures'. These were in use as supplements for the design of offshore structures in Norway for several years. Concrete structures were normally designed according to the Norwegian standard for concrete land-based structures of the day (namely NS3473).

In the Norwegian petroleum industry, the Ekofisk Bravo accident (1977) and the *Alexander L. Kielland* flotel accident (1980) were important for the first major steps for improved safety regulations. The *Alexander L. Kielland* flotel accident investigation (Moan 1981) led to several improvements in the Norwegian offshore oil and gas industry, most notably the requirements for performing fatigue analyses of semi-submersible platforms were introduced. However, several other elements also played an important role in the then ongoing improvement of safety in the offshore oil and gas industry, such as the 1977 regulation on labour safety, the introduction of regulation on 'Internal control' (the requirement for a responsible party to have in place a system to verify that they follow the rules and regulations).

The first mention of the annual exceedance probability of 10^{-4} for accidental situations was in a letter from NPD in 1979 – 'Guidelines Concerning Control of Limit State for Progressive Collapse, distributed by letter of 25 June 1979'. In 1981 this requirement was included in NPD's 'Guidelines for Safety Evaluation of Platform Conceptual Design', with the text 'the total probability of occurrence of each type of excluded situation should not by best available estimate exceed 10^{-4} per year for any of the main safety functions'. The structure would by the given definition normally be one of these safety functions. This regulation of accidental events led to significant research work on risk analysis and various accidental events that were relevant for structures (e.g. Moan 1981; Søreide et al. 1982).

In 1984 the NPD updated its regulation for 'Structural design of load-bearing structures intended for exploitation of petroleum resources' (NPD 1984). A load level with an annual probability of exceedance of 10^{-4} for checking structures in the limit state for progressive collapse was clearly included for environmental loads and accidental loads. A check of structures in damaged situations was also included. In 1985, NPD issued a risk-based functional regulation for all areas of its regulatory responsibility and were also given the regulatory responsibility for semi-submersible drilling rigs and flotels.

The limit state for progressive collapse was renamed the Accidental Limit State (ALS) in the 1992 version of the NPD regulation for 'Loadbearing structures' (NPD 1992). In addition, a risk analysis to evaluate the consequence of single failures was included. Structural reliability methods were mentioned but the use of structural reliability methods was rather limited. It was stated that the safety level (target probability) should be calibrated directly against the safety (failure probability) of known structural types and should be based on corresponding assumptions. In addition, structural reliability analysis should be proven to be on the safe side (conservative). Papers on how to perform such evaluations were published, e.g. by Moan (1993, 1998).

As stated in Moan (1983), the dominant hazard to structures used in the offshore oil and gas industry is the natural marine environment, the latent energy in hydrocarbons and human errors and omissions. The NPD regulation of 1992 to a large extent did take these into account by requiring structures to be designed according to the limit state method with partial factors. Risk analyses were required to determine, in particular, the accidental loads. Further, requirements were included with respect to damage tolerance.

The latest addition to the safety regulation of the offshore oil and gas industry has been the introduction of the principle of barriers (NPD 2002), see Section 2.2.3, and the transfer of specific structural guidelines from the NPD regulation to the NORSOK N-series of standards in the same year. The NPD regulation was in 2004 transferred to the Petroleum Safety Authority (PSA) Norway (PSA 2004).

2.1.4 ISO Standards

The development of an international standard for offshore steel fixed structures began in the early 1990s (following the development of the LRFD version of API RP2A) with the setting up of several committees with experts to develop and improve existing codes and standards. The development process took many years and initially an overarching standard was developed. This has been developed further into the present ISO 19900:2013 (ISO 2013b) which specifies general principles for the design and assessment of offshore structures subjected to known or foreseeable types of actions. These general principles are applicable to all types of offshore structures including bottom-founded structures as well as floating structures and for all types of materials used including steel and concrete. The principles are also applicable to the assessment or modification of existing structures. Later a draft standard for fixed offshore structures was issued for discussion. ISO 19902:2007 (ISO 2007) was finally published in 2007 and is recognised as the most contemporary international standard for the design and assessment of fixed offshore structures. Further ISO standards were developed for concrete structures (ISO 19903:2006; ISO 2006a), floating structures (ISO 19904: 2006; ISO 2006b) and site-specific assessment of jack-ups (ISO 19905:2016; ISO 2016). However, although these standards were not used in the early design of offshore structures, they are now the current standards against which both design and life extension assessments should be made. ISO 19902 includes a section on SIM which is discussed in more depth in Section 2.4.

2.2 Current Safety Principles Applicable to Structural Integrity

2.2.1 Introduction

A number of strategies have been published on safety principles. In general, safety principles linked to hazard management can be listed as:.

- Elimination (remove hazard).
- Substitution (prevent hazard).
- Engineering controls (isolate people from the hazard).
- Administrative controls (change the way the work is done to reduce likelihood of contact with the hazard).
- Personal protective equipment (protect people from being injured by the hazard).

In general, safety principles indicate that hazards should be identified and assessed and the need for protective measures against these hazards should be evaluated. Necessary requirements for the protective measures need to be established in order to ensure that they function as intended. This is often measured by their integrity, availability and robustness, which are important to be maintained during operation. For structures, this will include maintaining the strength, ductility and redundancy and having an organisation that can take care of the necessary SIM activities. Measuring performance and investigating incidents and accidents in order to identify improvement areas and improving the organisation ensures a continuous learning cycle.

Based on many years of experience of designing structures, a set of principles for safe design has been developed and embodied in codes and standards. A structure should be designed and fabricated in such a way that it satisfies these general ideas and principles during its intended service life:

- Having a structural configuration which has low sensitivity to the relevant hazards.
- Ability to sustain all actions and influences likely to occur during fabrication, erection, operation and demolition.
- Remain fit for the use for which it is required over its planned lifetime.
- Be designed to have adequate resistance, serviceability and durability.
- Avoidance as far as possible of structural systems that can collapse without warning.
- Selecting a structural form and design that is damage tolerant, e.g. so that it can survive adequately the accidental removal of an individual member or a limited part of the structure, due to local failure.
- Not be damaged by such events as explosion, impact and by the effect of human errors to an extent disproportionate to the original cause.
- In the event of an accidental event (e.g. fire), to provide load-bearing capacity for the required period for evacuation.

Ageing can affect the principles listed above in several ways. The extent to which the original design has met the current fitness for use and durability requirement will be evident and the structure may have experienced accidental events requiring repair. At this stage, the state of the structure may not meet the original design requirements, and reassessment is then required to assess the structure's current safety level and fitness for continued operation.

2.2.2 Application of Safety Principles to Structures

2.2.2.1 General

A safe structure that can withstand all load situations and accidental events at all times is not feasible. This is due to the uncertainty and inherent randomness in the strength of the structure, the loads and the accidental events. In addition, several aspects will not be foreseeable.

The strength of the structure, load situations and accidental events are not deterministic, predictable quantities. The strength of material varies, the quality of fabrication work varies, etc., and hence the strength of a structure also varies. Load situations are unpredictable and have an inherent randomness to them. Accidental situations can be defined but these may occur at a higher level or in a different way compared with what is predicted. Hence, it is typically not possible to foresee all accidental events the structure will be exposed to. Errors made in design, fabrication and the use of the structure are also impossible to foresee. In a few cases new knowledge about an unknown phenomenon has also led to the realisation that the structure was not designed correctly and with sufficient strength.

The traditional methods used to ensure that a structure is sufficiently safe (acknowledging that some structures may fail but with a very low probability) is to design them according to the following principles.

1. Strength according to the partial safety factor limit state design method (also called load and resistance factor design) is based on the following:
   - A characteristic value of material strength is used, which is a probabilistic defined low value of strength – typically in the range 2–5%. This is intended to ensure that there is a low probability of the strength being lower than what is assumed in the calculation.
   - Similarly, a characteristic high value of load is used – typically with an annual probability level of being exceeded of 10^{-2} for extreme loading situations and up to 10^{-4} for abnormal loading situations. This is intended to ensure a low probability of the load being greater than what is assumed in the calculation.
   - Risk assessment or standards are used to establish accidental events and accidental loads the structure may be exposed to. A characteristic high value of the loads is determined, typically at an annual probability level of 10^{-4} of being exceeded. Again, this is intended to ensure a very low probability of the accidental load being greater than what is assumed in the calculation.
   - The characteristic strength is reduced by a predefined safety factor into what is called design strength and the various types of characteristic loads are increased by individual/partial safety factors accounting for their assumed uncertainty. Higher safety factors (for uncertain loads such as wind, waves and earthquake) and lower safety factors (for less uncertain loads such as structural weight) are used for what are called design loads.
   - The structure is checked for predefined limit states (ULS, ALS, fatigue limit state [FLS] and serviceability limit state). Partial safety factors for strength and loads will vary for the different limit states but in general a limit state is a check that the strength is greater than the loads.
2. In addition to designing strength according to the partial safety method, a structure should also be sufficiently damage-tolerant to be able to withstand local failure without collapsing. This is meant to ensure some robustness for unforeseen

exceptional loads, unanticipated degradation, accidental events and unknown phenomena. Robustness is addressed in more depth below.
3. Structures are to be managed during their operation in order to uphold the integrity that they were designed to have.

2.2.2.2 Partial Factor and Limit State Design Method

The concept of limit states and partial safety factors as a design philosophy includes several independent safety factors. Each of these plays a particular role to ensure the safety of the structure against the exceedance of a limit state. The partial factors are of two major types:

1. Partial safety factor for a material and the soil, which takes into account the statistical variability of the strength properties for materials and the soil, fabrication and modelling of material parameters.
2. Partial safety factors for loads that take into account the possible deviation of the actual loads from the design (standard) values due to the variability of loading and departures from normal service conditions.

The various limit states, or conditions, that a structure needs to be able to withstand, due to applied actions during its life, are divided into two main groups:

- ULS, which is a failure check of the structure or one or more of its members due to fracture, rupture, instability, excessive inelastic deformation, etc.
- Serviceability limit states, which are a check of deflections and vibrations, etc.

The ULS includes failure modes of the structure such as:

- Excessive yielding (and possible rupture in non-linear analysis).
- Buckling due to elastic or elasto-plastic instability, leading to loss of equilibrium in part of or in the whole structure.
- Loss of equilibrium of the structure as a rigid body (overturning).
- Excessive deformation.
- Transforming the structure into a plastic mechanism/forming a mechanism.
- Geotechnical failure.
- Failure caused by fatigue or other time dependent effects.

The structure may fail in a ULS due to a single extreme load event or from a deterioration process over time followed by a milder load event. Exceedance of a ULS is almost always irreversible, and will cause permanent damage, deformation or failure.

Important sub-groups of the ULS are:

- ALS, which is a check of the collapse of the structure due to the same reasons as described for the ULS but exposed to abnormal and accidental loading situations.
- FLS, which is a check of the fatigue $S–N$ capacity or the crack growth capacity of the structure.

In the ALS, the effect of abnormal loading (e.g. very low probability environmental events) and possible accidental loads (such as collisions, explosions and fires) on the structural behaviour is considered. Many versions of ALS also include a check of the post-accidental condition where, for example, the structure is checked for representative loading situations after an accident such as a fire or explosion, with the purpose of ensuring that the structure will maintain its integrity to allow for escape and rescue

Table 2.1 NORSOK N-003 (Standard Norge 2017a) combination of annual probabilities of environmental loading.

Limit states		Wind	Waves	Current	Icing	Sea ice	Ice bergs	Snow	Earth-quake	Sea level
Ultimate limit states	1	10^{-2}	10^{-2}	10^{-1}	—	—	—	—	—	$HAT + S_{10^{-2}}$
	2	10^{-1}	10^{-1}	10^{-2}	—	—	—	—	—	$HAT + S_{10^{-2}}$
	3	10^{-1}	10^{-1}	10^{-1}	10^{-2}	—	—	10^{-1}	—	MWL
	4	10^{-1}	0.63	10^{-1}	—	10^{-2}	—	—	—	MWL
	5	10^{-1}	10^{-1}	10^{-1}	—	—	10^{-2}	—	—	MWL
	6	10^{-1}	10^{-1}	10^{-1}	10^{-1}	—	—	10^{-2}	—	MWL
	7	—	—	—	—	—	—	—	10^{-2}	MWL
Accidental limit states	1	10^{-4}	10^{-2}	10^{-1}	—	—	—	—	—	$MWL + S_{10^{-4}}$
	2	10^{-2}	10^{-4}	10^{-1}	—	—	—	—	—	$MWL + S_{10^{-4}}$
	3	10^{-1}	10^{-1}	10^{-4}	—	—	—	—	—	$MWL + S_{10^{-4}}$
	4	10^{-2}	10^{-1}	—	10^{-4}	—	—	—	—	MWL
	5	—	—	—	—	10^{-4}	—	—	—	MWL
	6	0.67	0.67	0.67	—	—	10^{-4}	—	—	MWL
	7	0.67	0.67	—	—	—	—	10^{-4}	—	MWL
	8	—	—	—	—	—	—	—	10^{-4}	MWL

HAT, highest astronomical tide; MWL, mean water level; S_q, storm surge at q annual probability.

before collapsing. This is, for example, the case in the Health and Safety Executive (HSE) safety case regulation (HSE 2015) and NORSOK N-series of standards (Standard Norge 2012).

An example of the annual probability level required for characteristic loads and their combination with other simultaneous loads in ultimate and accidental limit states is given in Table 2.1.

In most structural standards, wave loads need to be combined with wind and current loads, as indicated in Table 2.1. This is due to waves and wind being highly correlated and hence these are both required to be combined at an annual probability level of 10^{-2}. Current, however, is assumed to be less correlated with waves and hence is set to be combined with waves at an annual probability level of 10^{-1}.

The limit state and partial safety factor method is a so-called semi-probabilistic method. It takes into account that there is a chance that a structure becomes unfit for use, which in this context means that a particular limit state condition is exceeded. There is, however, no attempt to calculate that probability. The variable nature of any given parameter of the structural system (most generally strength and loads) is defined using statistics and a resulting 'characteristic' value is chosen for design calculations. For example, a characteristic load is defined as the load that has a certain chance of being exceeded at least once during the life of the structure (for example, a 10% characteristic dead load has a probability of 0.1 of being exceeded).

Having defined characteristic values of strength and loads, the design values for a particular limit state are the characteristic values of strength and load factored by the relevant individual partial safety factors. This procedure results in design values which

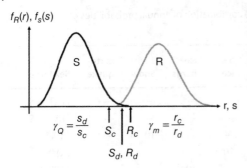

have a very low but unknown probability of being exceeded. The partial factors thus serve to deal empirically with the uncertain and extremely low probabilities associated with the tails of the probability distribution functions.

The general form of the limit state and partial factor method can be expressed as:

$$\phi R_c \geq \sum_{i=1}^{m} \gamma_i \cdot S_c$$

where ϕ is the strength factor, R_c is the characteristic strength, γ_i is the load factor of the ith load components out of m load components, and S_c is the characteristic value of the ith load component out of m components.

In many standards the general form is written with a material factor γ_m such that $\gamma_m = 1/\phi$. The limit state function then becomes:

$$R_c / \gamma_m \geq \sum_{i=1}^{m} \gamma_i \cdot S_c$$

Normally, this equation is written in the form of design value for load (S_d) and resistance (R_d):

$$R_d \geq \sum_{i=1}^{m} S_d$$

An illustration of the concept of partial factors (LRFD) is given in Figure 2.1 based on simple distribution functions for load and resistance.

The limit state and partial factor method is in most standards developed for the design of new structures. To some extent these standards may be used also for existing and ageing structures but this will often depend on careful inclusion of ageing effects beyond what was assumed in the writing of these standards. In practice, the effects of ageing, such as corrosion, cracking, denting, etc., have not been included in the design formulae, and the engineer is often left to rely on other information such as research reports and papers.

2.2.2.3 Robustness

As indicated earlier, history shows that structural failures and accidents occur due to a combination of:

- Limit state violation (stresses being higher than the strength, overturning moment being higher than the restoring moment, e.g. due to extreme weather, ship collision, etc.).

- Deterioration of the structure.
- Accidental events (e.g. fire, explosion).
- Human errors.
- Unknown phenomena.

Hence, a rigorous design process and material selection are important. However, a robust structure will in addition be able to withstand some damage, deterioration, and other changes that the structure may experience.

In relevant general standards for structures, e.g. ISO 2394:1998 (ISO 1998), ISO/DIS 2394:2013 (ISO 2013a), ISO 19900:2013 (ISO 2013b) and EN 1990:2002 (EN 2002), design principles[1] are given that indicate requirements for robustness. These are:

- Knowing and controlling the hazardous events and actions.
- Limiting the structure's sensitivity to the hazardous events and actions.
- Ensuring the structural elements are able to withstand the stresses from the experienced hazardous events and actions (limit state design) beyond the theoretical minimum.
- Ensuring that a single structural element failure is visible or detectable prior to complete collapse of the structure.
- Ensuring that the structure has the necessary damage tolerance.
- Reduction of the consequences of a collapse of the structure.

Part of a robust design is to reduce a structure's sensitivity to loads, accidental events and the environment it is operating in (e.g. a corrosive environment). An example would be to design structures with sufficient air gap (see Definitions) to avoid wave-in-deck loads on an offshore structure as most structures are very sensitive to loads onto the deck. Another example would be to use materials that are not sensitive to corrosive environments or applying a corrosion protection system that reduces the sensitivity to the corrosive environment. Designing for accidental and abnormal events will force the designer to evaluate how the structure will behave under more onerous loads (e.g. larger waves) and will ensure that the sensitivity to modified loads, accidental events and environment is as low as possible.

In some rare cases, proof loading can be used to check that a structure can tolerate the expected loads and will not fail from a limit state violation. For example, for pipelines this is a viable way of proving the strength. It is very difficult to apply to offshore structures.

Designing the structure with strength beyond the theoretical minimum will introduce some tolerance, e.g. by adding sacrificial thickness to allow for some corrosion and adding extra reinforcement (for concrete structures) for reassurance. Designing for accidental and abnormal actions is to some extent a way of making sure that the extra strength is placed where it is needed.

Damage tolerance in the form of several possible load paths, redundancy, ductility, etc. will be important for a structure if a single or limited number of failures occur. Damage tolerance should provide the necessary ability in the structure so that it does not collapse due to failure of one or a limited number of members and components.

Reducing the likelihood of failures due to unknown phenomena is of course difficult, for the obvious fact that they are unknown. Research and development is important in

---

1 It should be noted that all these principles are not given in each of the referred standards but this list of principles is an accumulation of the principles of safe structural design given in these different standards.

order to understand such phenomena as soon as they are identified. Keeping a close eye on the structures and other owners' experience with similar structures will help identify unknown phenomena at an early stage. Examples of such unknown phenomena in the offshore industry are the green water experiences on floating production, storage, and offloading units (FPSOs), wave slamming loads, etc.

2.2.2.4 Design Analysis Methods

Linear elastic analysis is the most used analysis method for determining the stresses in a structure due to the loading it is exposed to and hence performing a strength check based on the limit state method. In linear elastic Finite Element Analysis (FEA), the set of equations describing the structural behaviour is then $K \cdot r = R$. In this matrix equation, K is the stiffness matrix of the structure, r is the nodal displacements vector and R is the external nodal force vector. This equation is based on the assumption that displacements are small and can be neglected in equilibrium equations, the strain is proportional to the stress (linear Hookean material model), loads are conservative (independent of deformation) and the structural supports remain unchanged. Though the behaviour of real structures is non-linear, i.e. displacements are not proportional to the loads, non-linear behaviour may be neglected in most practical problems and linear analysis can be used. However, the behaviour of piles and foundations requires a non-linear representation to be included.

The fundamental requirement in structural analysis is that the calculations should be on the safe side. According to the lower bound theorem of plasticity, an external load in equilibrium with internal stresses which do not exceed the acceptable plastic stresses is less than or equal to the collapse load if the ductility is acceptable (Chakrabarty 1987). This may be fulfilled by using the above mentioned linear elastic analysis, giving statically admissible forces, followed by a code check of stresses according to accepted standards and the use of ductile materials. As this is in accordance with the lower bound theorem, it will in itself include a certain degree of safety towards collapse.

Non-linear analysis is an alternative to linear elastic analysis. In non-linear analysis, non-linearities such as geometric non-linearities (the effect of large displacements on the overall geometric configuration of the structure), material non-linearities (e.g. plastic material behaviour), boundary non-linearities (displacement dependent boundary conditions) and non-conservative loads (loads that are dependent on deformation) are included in the analysis. In the FEA the set of equations describing the structural behaviour is then $K(r) \cdot r = R(r)$, where both the stiffness matrix of the structure and the external nodal force vector are now dependent on the nodal displacements. The idea behind the non-linear analysis is to make the criteria for the maximum strength of the structure more realistic by better simulation of the real behaviour of the structure during collapse (Søreide 1981). Methods used for collapse analysis include geometric stiffness and material non-linearities. The methods will also give a solution according to the lower bound theorem. This solution will in most cases be more accurate than the linear elastic solution and closer to the theoretical collapse capacity.

Both linear elastic analysis and non-linear analysis follow the lower bound theorem in the theory of plasticity as the normal design principles for structures. Hence, both should provide a lower bound estimate of the structural strength.

However, in performing non-linear structural analysis one needs to be aware of the following issues:

- The principle of superposition cannot be applied. Thus, the results of several load cases cannot be combined and the results of the non-linear analysis cannot be scaled.
- Only one load case at a time can be handled.
- The sequence of application of loads (i.e. the loading history) may be important. In particular, plastic deformations depend on the manner of loading.
- The structural behaviour can be markedly non-proportional to the applied load.
- The initial state of stress (e.g. residual stresses from heat treatment, welding, etc.) may be important.

The theory behind the non-linear analysis of structures in general will not be covered here. The reader is referred to e.g. Crisfield (1996) and Skallerud and Amdahl (2002).

There are many challenges with respect to the use of non-linear analysis in ULS and ALS checks, however some standards give guidance on the use of non-linear analysis, especially for ALS checks. See e.g. NORSOK N-006 (Standard Norge 2015) and DNVGL RP-C208 (DNVGL 2016).

2.2.2.5 Management of Structures in Operation

Both the structure and its assessment will be changed in many ways during its operation. As indicated in Section 1.1, these changes will be of four main types:

- Physical changes (deterioration, damage, etc.).
- Technological changes (not often relevant for structures but very relevant for marine systems in floating structures and may be relevant if the structure is operated by control systems).
- Changes in the information and knowledge about the structure itself.
- Changes to knowledge in physical models, physical phenomena and engineering methods.

The integrity management of structures needs to ensure that such changes are identified, their impact evaluated and the changes mitigated if necessary in order to keep the structure safe.

An ageing structure will typically have accumulated many such changes and if these are not managed and mitigated properly during its operation it may be challenging and expensive to bring the structure back to a sufficiently safe state. However, in most cases structures should have been maintained properly, the necessary knowledge about the structure should be available and it should be possible to perform an evaluation of a structure's safety.

One of the most common failure causes for structures is due to design and fabrication errors, as indicated in Chapter 1. An ageing structure, especially one that has been exposed to loads close to the design load, will most likely have shown evidence of such errors. Hence, on the positive side an ageing structure may have a lower likelihood of having design and fabrication errors, since these would have been evident in earlier life.

2.2.3 Managing Safety

The UK's HSE document HSG 65 (HSE 2013) is a general guide to managing health and safety in the workplace. It is based on the principles of Plan, Do, Check, Act which achieves a balance between the systems and behavioural aspects of safety management.

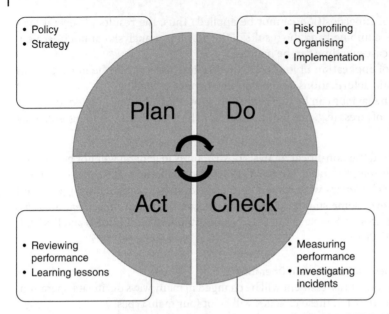

- Policy
- Strategy

- Risk profiling
- Organising
- Implementation

Plan | Do

Act | Check

- Reviewing performance
- Learning lessons

- Measuring performance
- Investigating incidents

Figure 2.2 Principles of integrity management. Source: Based on HSG65 (HSE 2013).

It also treats health and safety management as an integral part of good management generally, rather than as a stand-alone system. Its principles are applicable to structural safety management in a general way (Figure 2.2).

The Plan, Do, Check, Act management loop consists of several elements which include:

- *Plan.* Policy and strategy are a central requirement of this stage. Hazards should be assessed and the need for barriers to protect against these hazards should be evaluated. Performance standards should be established in order to ensure that the barriers are functional and have the necessary integrity and robustness to perform the task they are supposed to do. Procedures, inspection and emergency plans are required.
- *Do.* This entails operating in accordance with the strategy, risk profile and performance criteria. Further, maintain the strength, ductility and redundancy of the structure. Having an organisation that can take care of the necessary SIM activities. Data collection, evaluation, surveillance programmes and surveillance execution are part of this.
- *Check.* Measuring performance and investigating findings, incidents, and accidents in order to identify improvement areas in the structure and the SIM system.
- *Act.* Identifying ways to improve the structure and SIM based on the learning from existing structures, incidents and accidents.

Barriers are used in many different ways to manage risk in high hazard industries. One of the most detailed definitions of barriers is found in the regulation of the PSA Norway. In the PSA regulations, a safe and robust solution is assumed to be in place in order to avoid having hazardous situations but barriers should be in place to protect people, the environment and assets in case such a hazardous event should happen. Hence, a barrier is defined as a function that is intended either to prevent a chain of events (based on a

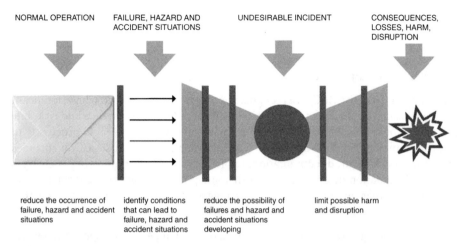

NORMAL OPERATION   FAILURE, HAZARD AND   UNDESIRABLE INCIDENT   CONSEQUENCES,
                   ACCIDENT SITUATIONS   LOSSES, HARM,
                                                                  DISRUPTION

reduce the occurrence of   identify conditions   reduce the possibility of   limit possible harm
failure, hazard and accident   that can lead to   failures and hazard and   and disruption
situations   failure, hazard and   accident situations
    accident situations   developing

Figure 2.3 A typical barrier diagram for handling situations outside normal operation – the design envelope (Vertical bars represent barriers). Source: based on Ersdal (2014).

hazard, failure, damage, etc.) from occurring or to affect a chain of events in such a way that limits harm and/or losses (Figure 2.3).

The functions of the barriers will typically be to identify the hazardous event, stop the hazard from escalating or limit the possible harm and inconveniences if the hazard escalates. A barrier function will be performed by technical systems or by organisations (people) and operations (actions). These are in the PSA regulations known as technical, operational and organisational barrier elements.

A barrier element will typically have a function in the normal operation of a facility as well as being a barrier element. A few examples of technological systems and equipment that are merely functioning as barrier elements can be found (e.g. in cars, seat belts and airbags do not have any function in the normal operation of the car but are clearly an element in a barrier with a function to limit harm to people in hazardous situations such as crashes). Some operations (actions) are primarily performed in hazardous situations as such, merely barrier elements. It should be noted that the same personnel are used both in normal operation and in hazardous situations.

The primary function of an offshore structure is to support the topside with its equipment and personnel in the normal operation. However, an offshore structure is also expected to withstand the loads in some hazardous situations (e.g. ship impact) and is, as such, a barrier element in some hazardous scenarios. The structure is also expected to withstand fire for a limited amount of time and to withstand explosions up to a certain level. In addition, the structure is required to not collapse in the event of local damage occurring due to an identified or unidentified accidental event. In structural standards, these hazards are covered in the ALS assessment.

The most important addition a barrier approach will add to structural engineering is systematic thinking around the hazards that may occur and how to design and protect the structure against these hazards. Such mitigating actions may be the removal of personnel prior to a storm, or to limit the weight that is allowed on the platform, etc.

Ageing barrier elements and structures will degrade and a resulting hazard may arise from failures of the barrier elements and the structure itself. In addition, the barriers and structures may be less able to resist accidental actions. Also, the information and knowledge about the operational limits may be no longer available.

2.2.4 Change Management

Management of Change (MoC) is a recognised process that is required when significant changes are made to an activity or process which can affect performance and risk. Life extension falls into this category and the primary objective of an MoC process is to ensure that sufficient rigour is applied to asset ageing and life extension. This includes the planning, assessment, documentation, implementation and monitoring of changes affecting an installation or operation such that any potentially adverse effects of or on ageing and life extension are identified and managed effectively. Particular aspects that need consideration in the MoC process are as follows:

- Recognising ageing and obsolescence as a trigger for change at any stage in the service life of an asset.
- Any modifications, upgrades and repairs.
- Changing standards.
- Process operating conditions and safe operating limits.

Special consideration is required for managing the change or transition from the originally anticipated design life to a longer life period. It is necessary to demonstrate that where assets are expected to be operated beyond their anticipated service life the asset has sufficient integrity and remains or is expected to remain fit for purpose for the specified life extension period. Asset life extension planning is required as part of the MoC process.

2.3 Current Regulation and Requirements for Ageing and Life Extension

2.3.1 Regulatory Practice in the UK for Ageing and Life Extension

The regulatory requirements for the SIM of structures operated on the UK Continental Shelf (UKCS) are specified in the following (current in 2017):

- The Safety Case Regulations (HSE 2015), which make preparation of a safety case a formal requirement.
- The Design and Construction Regulations (DCRs) (HSE 1996), which require the duty holder to ensure that suitable arrangements are in place for maintaining the integrity of the installation, through periodic assessments and carrying out any remedial work in the event of damage or deterioration. This regulation introduced the concept of safety critical elements (SCE's) defined as those parts of an installation the failure of which could cause or contribute substantially to, or the purpose of which is to prevent or limit the effect of a major accident. Typical structural SCEs are the platform sub-structure and the topsides.

The duty holder needs to provide evidence that foreseeable structural damage to the installation, escalation potentials and all likely scenarios have been considered.

Thus, it is of utmost importance that deterioration and degradation are incorporated into a well-formed SIM system and associated plan.

The revision of the Safety Case Regulations in 2005 (HSE 2005), and retained in the latest revision of 2015 (HSE 2015), explicitly addressed for the first time the extension of operation beyond the original design life. As part of the update of the UK Safety Case legislation in 2005 (HSE 2005), a revised safety case is required to be submitted to the HSE where material changes to the previous safety case have occurred. These include:

- Extension of use of the installation beyond its original design life.
- Modifications or repair to the structure where the changes have or may have a negative impact on safety.
- Introduction of new activities on the installation or in connection with it.
- Decommissioning a production installation.
- Thorough review.

The original design life was usually based on fatigue criteria and in the absence of a defined life the recommendation was to assume a design life of 20 years (HSE 1990). A revised safety case would need to identify all of the hazards with the potential to cause a major accident and how the major hazards arising from these are or will be adequately controlled. These hazards include those arising from ageing processes associated with life extension, such as corrosion and fatigue. The UK safety case legislation requires reassessment of safety cases every five years by the duty holder, so in terms of life extension a case for an extended life of five years is the maximum requirement.

The latest version of the UK Offshore Safety Case Regulations (HSE 2015) introduced the concept of Safety and Environmental Critical Elements (SECEs) which was extended to cover not only major accident hazards but also elements the failure of which could lead to significant environmental damage. This concept was introduced as a result of a European Directive (EU 2013).

For installations on the UKCS there is a formal requirement as part of the safety case legislation for a verification scheme to be provided to the duty holder by which an independent and competent person (ICP) reviews the record of SCEs (which would include the platform's structure) and to provide any reservations on this. The verification scheme needs to be reviewed on a regular basis and, where necessary, revised or replaced in consultation with the ICP. This verification process should include the review of performance standards which are normally associated with SCEs. The performance standards are normally based on four main elements: (i) functionality; (ii) availability and reliability; (iii) survivability; and (iv) dependency. In terms of life extension, the important elements are availability and reliability and survivability, all of which could be affected by ageing processes such as fatigue and corrosion. It is noted that the verification scheme provides information primarily to the duty holder and not to the regulator (to which it could provide some benefits).

To date, structural performance standards are less well developed than those for accidental hazards such as fire and explosion. The application of performance standards to offshore structural components has been addressed in HSE (2007). It was noted that design criteria in codes and standards provide a strong basis for setting structural performance standards. In addition, inspection and maintenance are recognised as important for maintaining performance standards, particularly when ageing effects are becoming important. Key performance indicators for offshore structural integrity have been developed and are reported in Sharp et al. (2015).

Audit is also a formal regularity requirement, as part of the safety case regulations. Such audits are important in demonstrating the organisation's ability to follow the processes which have been set in place. In terms of SIM, the development of an overall inspection philosophy and strategy and criteria for in-service inspection are important and would be expected to be demonstrated as part of an audit. The demonstration of structural fitness for purpose at the evaluation stage is also very important for the overall safe performance of an installation and should be checked as part of the audit function.

Use of modern codes and standards is recommended by HSE in achieving good practice, which is a requirement in the ALARP approach (HSE 2015).

Recommendations for ageing semi-submersibles are given in HSE (2007). It is stated that duty holders should periodically reassess their arrangements that are used to maintain integrity to take account of the effects of ageing processes and to ensure that any deterioration is detected in good time, particularly for installations beyond their notional design life. A number of measures should be considered (if not already in place) when reassessing the arrangements. These include fatigue life assessment both in the intact and damaged condition, inspection requirements, use of information on past performance, replacement/repair of ageing components, review of the effectiveness and reliability of barriers and review of the effect of changes in knowledge concerning technology and environmental conditions that could influence existing barriers or make further measures reasonably practicable.

2.3.2 Regulatory Practice in Norway Regarding Life Extension

Petroleum activity in Norway is regulated by the PSA (Petroleum Safety Authority) based on several regulations. The overarching regulation is the Framework Regulation giving the general principles. The Management Regulation deals with more detailed principles for risk management of petroleum activities. The Facility Regulation gives requirements to the design and layout of primarily new installations. Finally, the Activity Regulation gives requirements for the operation phase, e.g. including asset integrity management and SIM.

In Norway, there is a formal regularity requirement for an operator to obtain permission from the PSA to operate beyond the original design life (PSA 2016). In the PSA guidelines a list of requirements for the consent is provided. These include that an application for consent for life extension for a permanently placed platform should contain a summary of the operator's barrier management. This should include identification of needs for updated performance requirements, which take into account the fact that ageing effects can lead to the impairment of several barriers at the same time. Furthermore, the application should contain an assessment of potential preventive measures as well as:

1. An overview of non-conformities and gaps and how these are handled with regard to risk reduction.
2. A description of the operator's use of information regarding previous behaviour and use of relevant equipment, including experience from similar facilities. This can require cooperation with other operators, shipowners, and classification societies.

3. A description of the period the facility is planned to be used, identification of the factors that will limit life of the platform and an indication of criteria for safe operation to the extent possible.
4. The operator's plans for modifications, replacements and repairs, if required.
5. A description of changes in maintenance philosophy, strategy and programme, which will be implemented as a consequence of the expected ageing effects.
6. The period of time for which consent is applied.

The summary mentioned above should, according to PSA, be prepared in accordance with the Norwegian Oil and Gas Association's Guideline 122 (NOROG 2017), complete with supplementary standards, and should contain a résumé of analyses carried out according to this guideline. For structures and maritime systems, the summary should contain a résumé of analyses carried out according to NORSOK N-006.

In addition to these specific requirements to the life extension in itself, the Norwegian regulations have general requirements for SIM (PSA 2016). The requirement is that the facilities (installations including structures) should at all times be able to perform their intended functions. The functions of a structure must be assumed to include them to be sufficiently safe. The regulation on integrity management refers to NORSOK N-005 (Standard Norge 2017b). The Norwegian regulations also require performance standards to be developed for barriers. These standards are usually defined in terms of their functionality, integrity and robustness. It is expected that these performance standards should take account of ageing effects and that mitigating actions are taken to compensate for degraded barriers.

The Norwegian Maritime Authority (NMA 2016) regulates Norwegian flagged Mobile Offshore Units. As regards mobile facilities registered in a national ships' register operating on the Norwegian Continental Shelf (NCS), relevant technical requirements in the NMA's regulations for mobile facilities apply, with the specifications and limitations that follow from Section 1 of the PSA Facilities Regulations (PSA 2016). The PSA is issuing a letter of compliance for those flagged rigs that are to be used in petroleum activities. The PSA also gives permission for the use of rigs at site-specific locations.

A specific letter to the owners of ageing mobile offshore units was issued from the NPD (now the PSA) in 2003. This letter had among other requirements:

- Fatigue life calculated according to current rules and regulations and corrected for changes in assumed weights and weight distribution caused by modifications or changes in assumed usage.
- Verification of physical match between the facility and as-built documentation so that later modifications or changed usage are taken into account in analyses and calculations.
- Operators' additional considerations and requirements with respect to inspection and maintenance as a result of extended life for load-carrying structures with respect to fatigue, corrosion, erosion and thickness measurements.

2.3.3 Regulatory Practice in the USA

The US Government began regulating the offshore energy industry in the late 1940s but its jurisdiction was not fully established until the passage of the Outer Continental

Shelf Lands Act (OCSLA) of 1953. From 1982, regulation of activities on the Outer Continental Shelf was in the hands of the Mineral Management Services (MMS) which was an agency of the US Department of Interior which managed the nation's natural gas oil and mineral resources on the Outer Continental Shelf. After the *Deepwater Horizon* tragedy in April 2010, the BSEE (Bureau of Safety and Environmental Enforcement) was established in 2011 to separate regulatory responsibilities from activities concerning lease sales and revenue generation. MMS and now BSEE have relied on API Recommended Practices (RPs). Since the offshore activity in the Gulf of Mexico started much earlier than in the North Sea, the number of old platforms is much larger in the Gulf of Mexico. More strict design requirements have been introduced over the years and hence older platforms have a much lower capacity against environmental loads than new installations even without considering the effects of physical ageing processes such as corrosion and fatigue.

API has been the leader in developing industry standards that promote reliability and safety in the workplace. Over many years, API has produced a series of RPs covering many aspects relating to offshore safety and integrity. The Recommended Practice for Planning, Designing, and Constructing Fixed Offshore Platforms (RP 2A) was first produced in 1969 (API 1969). A new Section 17, 'Assessment of Existing Platforms', was added in the 21st edition (API 2000) as a result of Hurricane Andrew that led to the collapse or severe damage of a number of platforms in the Gulf of Mexico (PMB Engineering 1993). The damage caused by Hurricane Andrew showed the need for the offshore industry to standardise the documentation of the structural integrity of existing structures.

A new API RP (API RP 2SIM) was introduced in 2014 that deals with SIM of existing fixed platform structures (API 2014b). This RP incorporates and expands on the recommendations of API RP 2A Section 17. Although life extension is not listed as a trigger for structural assessment, a number of factors are listed which could apply to ageing installations. Specific guidance is provided for the evaluation of structural damage, above- and below-water structural inspection, fitness for purpose assessment, risk reduction and mitigation planning. For planning, designing, and constructing new offshore floating production systems guidelines, RPs and other requirements relating to including reuse and change in use of existing floating production systems are given in API RP 2FPS (API 2011).

2.3.4 Regulatory Practice Elsewhere in the World

As noted in Section 1.2, offshore development has taken place in several different parts of the world, leading to the development of national regulations. Many of these have been based on existing practice. In Australia, where offshore development began on the west coast, the National Offshore Petroleum Safety and Environmental Management Authority (NOPSEMA) is the body with responsibility for safety and environmental protection. The safety regulations are based on those implemented in the UK, with hazard management and a safety case as the basis for safety.

Both Denmark and the Netherlands have offshore structures dating back to the 1980s. In the Netherlands the relevant safety authority is the State Supervision of Mines and in Denmark it is the Danish Working Environmental Authority. Offshore safety regulations generally follow the risk reduction, safety management approach.

In Canada the regularity authorities are the National Energy Board (for frontier areas), the Newfoundland and Labrador Offshore Petroleum Board and the Nova Scotia Offshore Petroleum Board. Safety is based on a combination of goal setting and prescriptive approaches, including the need for a certificate of fitness to be obtained from one of several Certification authorities.

In Mexico the relevant safety organisation is the National Agency for Safety, Energy and Environment of Mexico (ASEA). ASEA requires that activities in the hydrocarbon sector are performed under Safety and Environmental Management Systems (SEMS).

In Brazil, where offshore development started mainly in the Campos basin, ANP (*Agência Nacional do Petróleo* – Brazilian National Regulatory Authority for oil, gas, and biofuels) has the responsibility for both exploration and safety.

In many of these areas, assessment of existing and ageing structures will be performed according to ISO standards, especially ISO (2007) and ISO (2017), but NORSOK standards, UK regulation and API standards are also used.

2.4 Structural Integrity Management

If everyone were clothed with integrity, if every heart were just, frank, kindly, the other virtues would be well-nigh useless

Moliere (Jean Baptiste Poquelin)

2.4.1 Introduction

In engineering terms, integrity is defined *as the state of being whole and undivided, the condition of being unified or sound in construction* (Oxford dictionary). SIM is a key process in ensuring the safety of offshore structures. The purpose of SIM is to identify changes, to evaluate the impact of these changes, mitigate the impact of these changes if found necessary with the aim of keeping the structure sufficiently safe during operation and use. As mentioned earlier, such changes may be physical changes, technological changes, changes to knowledge and safety requirements, and structural information changes. The physical changes will often be the most important and will normally dominate the SIM work. These will often include rather costly offshore structural inspections, structural monitoring, weight monitoring and metocean observations. However, the other types of changes should not be overlooked. These will often include document review, keeping updated on engineering methods and standards and database maintenance.

SIM is a subset of asset integrity management. Asset integrity management is about keeping the asset unimpaired and in sound condition throughout the lifecycle, whilst protecting health, safety, and the environment. However, in broader terms, asset integrity management is about caring for more than the physical assets. In PAS 55 (BSI 2008), for example, the assets that are meant to be managed include:

- Physical assets (plants, structures, equipment, etc.).
- Human assets.
- Information assets.
- Financial assets.
- Intangible assets (reputation, moral, intellectual property, goodwill, etc.).

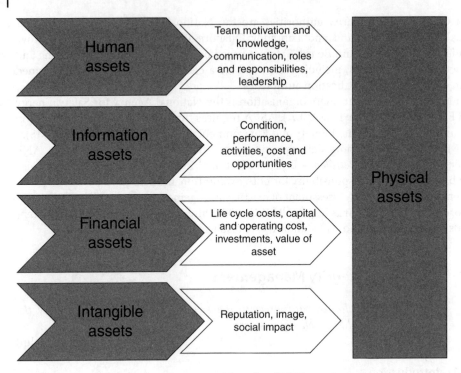

Figure 2.4 The relevant assets to be managed, based on PAS 55 asset integrity management. Source: based on BSI (2008).

In the context of this book, the focus is on the physical assets, namely the structures. However, structures cannot be looked at in isolation. They all interface with the physical assets in important ways (BSI 2008), see Figure 2.4.

Hence, asset integrity management is the means of ensuring that the people, systems, processes, and resources that deliver integrity are in place, in use and will perform when required over the whole lifecycle of the asset, including life extension if enabled (PAS 55). Asset integrity management can further be described as the continuous assessment process applied throughout design, construction, installation, and operations to assure that the facilities are and remain fit for purpose. The integrity management process covers the physical asset and those other systems that prevent, detect, control, or mitigate against a major accident hazard. A loss of integrity could have an adverse impact on the safety of personnel, on the safety of the asset, on the environment, or on production and revenue.

The aim of the asset integrity management process is to provide a framework for the following:

- Compliance with company standards, regulatory and legislative requirements.
- Assurance of technical integrity by the application of risk based or risk informed engineering principles and techniques.
- Delivery of the required safety, environmental and operational performance.
- Optimization of the activities and the resources required to operate the facilities whilst maintaining system integrity.
- Assurance of the facilities' fitness for purpose.

Figure 2.5 Asset management system. Source: based on IOGP (2008).

Several systems are in use to manage assets effectively and contain several elements which are needed for a successful outcome. These include the already mentioned PAS 55 system and the international version ISO 55001 (ISO 2014). In addition, the International Association of Oil & Gas Producers (IOGP) has issued its own document entitled 'Asset Integrity – the Key to Managing Major Incident Risks' (IOGP 2008). In the IOGP document, asset integrity is focused on the prevention of major incidents as an outcome of good design, construction and operating practices. It is achieved when facilities are structurally and mechanically sound and processes are performed as they were designed. It is based on a standard continual improvement cycle of Plan, Do, Check, Act (as described in Section 2.2.3). The key elements of the process described in the IOGP document are shown in Figure 2.5.

2.4.2 The Main Process of Structural Integrity Management

The aim of the integrity management of structures and marine systems is to ensure that an adequate level of safety and availability is maintained throughout the lifecycle. The SIM process, hence, includes the monitoring of relevant parameters, such as:

- The physical condition of the structure.
- The configuration of equipment, weight and the structure itself.
- The structural loading and the environment that the materials are exposed to.
- The knowledge of the structure and requirements for the maintenance of integrity.

These activities are supported, as appropriate, by structural assessments and the implementation of mitigating measures to control the likelihood of structural failure. In order to perform these tasks an organisation, a strategy and plan are needed and hence a process to achieve these. Guidance on the necessary SIM organisation, responsibilities of individuals and management processes is given in HSG 65 (HSE 2013), ISO 19902 (ISO 2007), ISO 19901-9 (ISO 2017), API RP 2SIM (API 2014b), and NORSOK N-005 (Standard Norge 2017b).

Offshore structures are exposed to severe environmental and external physical forces during operation which can lead to deterioration and the loss of integrity. This necessitates the formulation and implementation of maintenance strategies to ensure that structural integrity is retained at all times. Traditionally, the SIM process has entailed inspection and repair focused mainly on deterioration processes to ensure that the structure remained fit for purpose. However, the integrity management of a structure in operation in the present day includes a number of other activities contained within a wide framework in which inspection and repair are components.

The SIM process is often defined as: 'the collection of necessary information about the structure, its condition, its loadings and its environment to enable sufficient understanding of the performance of the structure to ensure that loading limits are not exceeded and that safe operation is assured'.

Information on the integrity of the structure is obtained by conducting surveys at different levels, including:

- Condition survey, which is to a large extent traditional inspection and monitoring of the structure for cracks, corrosion, damages, and other changes to the structure itself.
- Physical survey. Survey of issues such as sea bed (scour), marine growth and subsidence.
- Survey of the metrological and oceanographic environment (waves, wind, current, etc.) and other relevant loads.
- Survey of topside weight, operational loads and hazard conditions (live load management, visiting ship size and shape, mooring tension, fire and explosion scenarios, passing traffic routes, etc.).
- Survey of technological developments and improved understanding of failure mechanisms that may influence the safety of the structure.
- Survey of updates in standards, regulations, and other requirements.

Activities that are important in order to keep a structure safe in operation will typically be:

- Establishing and updating the external context: this includes the design basis, regulations and standards that were used when the structure was designed which are part of the context and assumptions that form the basis for the safety of the structure.
- Operating in accordance with limits and operational restrictions – establishing the necessary procedures that go together with the intent of the thinking behind the design of the platform – such as ballasting procedures, competence of marine personnel, etc.
- Maintaining data and information about the structure, including past inspections and surveys.
- Engineering evaluations of the impact of any findings, changes, etc.
- More detailed assessments of structures may be required if the findings are of concern (e.g. triggering assessments in accordance with the relevant standards such as ISO 19902).

- Execution of inspections, maintenance, repair and other survey programmes.
- Emergency preparedness (see Section 2.4.5) and response plans in the case of findings that influence the immediate safety of the structure.
- Evaluation of the activities in SIM in order to identify improvement areas.
- QA/QC of the activities related to SIM.

2.4.3 Evolution of Structural Integrity Management

2.4.3.1 The Early Years

In the 1970s and most of the 1980s, offshore structures in the petroleum industry were designed on the basis of evolving codes and standards. In the UK and Norway, early designs were influenced by the practice in the Gulf of Mexico. It was found from experience that the US practice did not properly account for fatigue which was much more damaging in the North Sea and several early repairs were required due to fatigue cracks developing early in the life.

Two key structural accidents, *Sea Gem* in the UK sector and *Alexander L. Kielland* in Norwegian waters, required regulators to review practice and implement improvements. The inspection plans were often based on routine intervals as found in ship rules and regulations and the parts of the structures that were regarded as major or important were inspected typically every fifth year. Traditional inspection strategies were costly and did not necessarily target the most critical structure.

The offshore industry began to consider more cost-effective and optimised inspection strategies. In the late 1980s, methods for probabilistic structural analysis or structural reliability analysis had developed to such an extent that they were used to determine the need for inspection in a risk-based approach. The idea was that the probability of a crack in a member or joint could be calculated by structural reliability analysis and that the consequence of failure of that specific member or joint could be calculated by so-called push over or collapse analysis of the structure with that member or joint removed. More details on these approaches are given in Section 4.4.4 and Section 4.7.

2.4.3.2 The Introduction of Structural Integrity Management into Standards

The development of the international standard ISO 19902 commenced in 1993 and included a section on SIM in early drafts. This standard was first issued in 2007 but the early drafts were used as the basis for SIM in the interim.

SIM in this standard took into account all the issues and methods that had been used to inspect, monitor, evaluate and assess fixed offshore steel structures over the years. Four major elements of SIM were identified:

- Data management.
- Engineering evaluation of structural integrity.
- Inspection strategy.
- Inspection programme (the execution of inspections).

In addition, assessment was introduced as an option following engineering evaluation (which is a more qualitative review). The elements of SIM are described in Figure 2.6.

2.4.4 Current SIM Approach

More recently, the offshore industry has adopted a major hazard approach to the integrity management of structures and other assets, e.g. NORSOK N-005 and

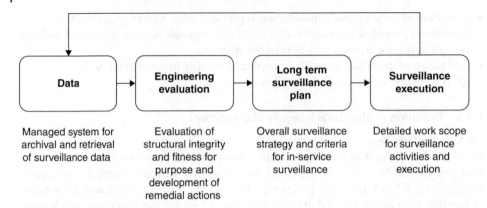

Figure 2.6 Structural integrity management cycle (surveillance includes inspection of the condition, determination of loading, review of documents, etc. needed to determine changes that may affect the safety of the structure). Source: based on ISO 19902.

ISO 19901-9. The major accident hazard based integrity management relies on typical risk analysis methodology and seeks to understand the hazards the structure is exposed to and to make sure that such hazards do not result in major accidents.

The basis of this current approach to SIM is defined, for example, in the UK's and Norway's regulatory regimes. The regulations are aimed at controlling the risk associated with major hazards. The risk control process entails the identification of the major hazards, the determination of the unwanted events that may result, investigation of the possible consequences from these events and that suitable barriers to protect against the hazards turning into unfavourable consequences are established and maintained throughout the life of the facility.

Specifically, this includes the determination of:

- The goal (safe structures, withstanding the loads to a 10^{-4} yearly probability level, intact functionality, etc.).
- The unwanted events that may hinder meeting the goals (e.g. global structural failure/collapse, extensive deformations or damage to the structure limiting the functionality of the facility, structural parts falling off the structure, etc.).
- The hazards that can cause such unwanted events.
- The consequences from these events.
- What hinders the hazard from causing the unwanted event or that is causing a preferred outcome (barrier or SCEs).

A key element of the major hazard and barrier management regulatory regimes is the prioritisation of activities, including inspection, maintenance and repair, for the important barriers in the system.

Several standards and reports have been published on the principles of SIM, e.g. HSE RR684 (HSE 2009). These identify a number of key processes which are considered good practice in SIM, together with an appropriate management and documentation structure.

A modern SIM system would typically include processes, such as:

- *SIM policy.* The SIM policy sets out the overall intention and direction of the duty holder with respect to SIM and the framework for control of the SIM related processes and activities.
- *SIM strategy.* The SIM strategy sets out the duty holder's plan for the integrity management of its assets in line with the SIM policy and sets acceptance criteria.
- *Surveillance strategy.* A systematic approach to the development of a plan for the identification of changes (e.g. in-service inspection of a structure to identify deterioration, identification of loading changes, document review to identify changes in standards and regulations, document review to identify changes in engineering methods, etc.).
- *Surveillance programme.* The surveillance programme, developed from the surveillance strategy, is the detailed scope of work for the surveillance to identify changes. This will typically include:
  - *Inspection programme.* A detailed scope of work for the offshore execution of the inspection activities to determine the current condition and configuration of the structure (see Section 5.2).
  - *Load assessment.* A detailed scope of work for the execution of activities necessary for identification of changes to loading (weight monitoring, metocean monitoring and updates of metocean reports, etc.).
  - *Document review.* A scope of work for the execution of the necessary document review in order to identity changes in engineering methods, knowledge about loadings and strength, standards, and requirements.
- *Structural evaluation.* Review of the current condition of the structure compared with when it was last assessed and other parameters that affect the integrity and risk levels to confirm or otherwise that the acceptance criteria for structural integrity are met. This process is expected to identify repair or maintenance needed to meet the acceptance criteria for structural integrity.
- *Repair and modifications.* The necessary mitigation activities in order to retain the safety of the structure, typically based on output from the structural evaluation.
- *Information management.* The process by which all relevant historical and operational documents, data and information are collected, communicated and stored
- *Audit and review.* Audit is the process to confirm that SIM is carried out in conformity with the procedures set out in the SIM policy and strategy and legislation. The review process assesses how the SIM processes can be improved on the basis of in-house and external experience and industry best practice.

A possible flow chart for the SIM process is shown in Figure 2.7, based on NORSOK N-005 and ISO 19901-9. Integrity assessment steps depend on initiators often given in standards, such as life extension. The compensating measures are key to the successful management of life extension and could include strengthening, clamps, repairs, hammer peening and several other methods for extending the life described in this book.

These standards and documents also give requirements for the competency of personnel involved in the SIM process, e.g. ISO 19902 (ISO 2007) and API RP 2SIM

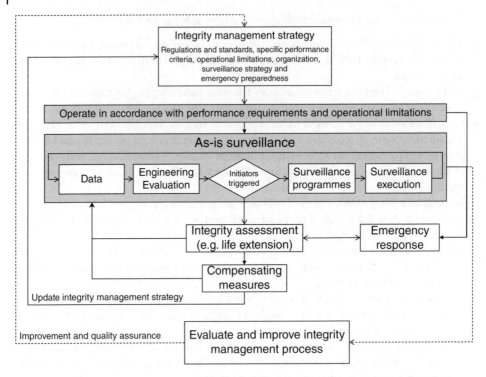

Figure 2.7 A structural integrity management process flow chart.

(API 2014b). In these standards it is stated that the engineer or group of engineers involved with SIM should be:

- Familiar with relevant information about the specific platform(s) under consideration.
- Knowledgeable about the underwater corrosion process and prevention.
- Experienced in offshore structural engineering.
- Experienced in offshore inspection planning.
- Knowledgeable in the use of inspection tools and techniques.
- Aware of general inspection issues in the offshore industry.

In addition, competency for inspectors is included in schemes such as CSWIP (2018), NOROG and NORDTEST (EN 473:2008; EN 2008). In many instances, competency management extends to external suppliers who may have a direct effect on SIM. These standards also give a competency requirement on understanding the difference between the design of new structures and assessing existing ones. The design of new structures is normally done according to present codes and standards with no actual information on how the structure is behaving when installed. Existing structures, however, should be assessed by the present standards with consideration of the original standard used in design and that actual performance data may be available (measurements, inspections, tests, and occasionally even a proof-load test).

2.4.5 Incident Response and Emergency Preparedness

Damage and failure of structural or marine systems may escalate rapidly and, hence, the response to such damage and failures needs to be treated with an appropriate urgency. Incidents such as loss of watertight integrity, unintended flooding of compartments, loss of mooring lines and dynamic positioning (DP) failures are examples of incidents that may require an urgent response. Loss of structural integrity may also lead to such situations if the damage or failure is to an area or member with little redundancy, i.e. where there is no alternative load path and the potential collapse of the structure is possible.

It is recommended that such situations are treated as emergencies and that operational personnel have clear and concise guidance that allow them to make the appropriate decisions in response to the loss of integrity.

Regulations (HSE 2015; PSA 2016) normally require the operator to have sufficient emergency preparedness and emergency response plans for reasonable foreseeable situations that may occur in normal and in severe conditions. However, few standards for structural integrity and marine system integrity give detailed further guidance. An exception to this is the proposed API RP 2MIM (API 2018), where annex D is devoted to incident response planning and emergency response for mooring failures.

The key elements of emergency preparedness relevant for structural and marine system engineers are:

- Knowing which structural and marine system situations that need to be addressed by emergency preparedness routines/procedures.
- Informing the organisation (specialised emergency-response-recovery organisation offshore and onshore).
- Awareness of potential emergency situations and their impact on structures and marine systems, if any.

Understanding the significance of potential emergency situations on the structure and marine systems is a key factor and preparing for these potential emergency situations is recognised as being important. The most important role of a SIM engineer in emergencies will normally be to support the response team with necessary structural and marine system competence, evaluations, etc.. Examples of possible hazards and accidental situations that may need structural and marine system competence assistance may be:

- Fire.
- Explosion.
- Ship collision.
- Adverse weather.
- Failure in stability (leakage, astray water, unexpected flooding, etc.).
- Failure in position keeping.
- Failure in critical structural elements.
- Failure in piles and soil support.

It is unrealistic to prepare for all types of hazardous situations related to structures and marine systems. However, being prepared for a set of known cases and having a competent support team with specialists on structures and marine systems is vital in order to provide the emergency response and recovery management with the best available information for making good decisions.

2.4.6 SIM in Life Extension

SIM in life extension has been handled historically as part of the ongoing maintenance routine for operational installations without formal recognition as an explicit activity. Lately, after initiatives of regulators such as HSE (UK) and PSA (Norway), more attention has been placed on structures in life extension.

It is normally required in standards or by regulation that foreseeable structural damage, escalation potentials and all likely scenarios have been considered. This would require the identification of deterioration and degradation to be incorporated into a SIM system and associated strategy.

The principles of a SIM strategy relating to life extension are presented in Table 2.2.

An important requirement in the structural integrity assessment of ageing installations and life extension is the availability of detailed information from inspections both at the fabrication stage and during the operational phase. However, information on the full inspection history is not always available. Inspection of structures is further dealt with in Chapter 5.

Structural evaluation is an ongoing process to confirm that the basis for demonstrating the structural integrity and associated risk levels is still valid. The results of this evaluation and possible additional assessments, and the effects of the subsequent control measures, need to be considered and used in updating the SIM strategy.

SIM requires that large amounts of information are collected and stored. To this effect, owners typically have computerised systems in place. However, data for older platforms are not always available, e.g. following changes in ownership. This lack of data requires careful treatment at the evaluation stage and possibly the use of higher than normal safety factors in the analysis.

Table 2.2 SIM processes and associated issues affecting life extension.

SIM process	Main issues affecting life extension
SIM strategy	The strategy should include managing the approach to assessing ageing processes and the need to link surveillance and inspection requirements to these
Surveillance programme	More detailed surveillance and inspection may be required if a period of life extension is to be justified
Structural evaluation	The evaluation should include assessment taking account of the original design requirement (which may have been less onerous than modern standards), as well as the consequences of ageing processes (e.g. fatigue, corrosion)
Information management	This may be influenced by loss of key data from original design, construction, installation and early operational inspections

Bibliographic Notes

Sections 2.2.1 and 2.2.2 are based on Ersdal (2014). Section 2.2.3 is partly based on HSG 65 (HSE 2013) and partly based on Ersdal (2017).

References

Adams, J.R. (1967). Inquiry into the Causes of the Accident to the Drilling Rig Sea Gem. The Ministry of Power, HMSO, London.

API. (1969). RP 2A Recommended Practice for Planning, design and constructing fixed offshore platforms. API Recommended Practice 2A, 1e. American Petroleum Institute.

API. (1972). API RP 2A Recommended Practice for Planning, design and constructing fixed offshore platforms. API Recommended Practice 2A, 4e. American Petroleum Institute.

API. (1977). API RP 2A Recommended Practice for Planning, design and constructing fixed offshore platforms. API Recommended Practice 2A, 9e. American Petroleum Institute.

API. (1982). API RP 2A Recommended Practice for Planning, design and constructing fixed offshore platforms. API Recommended Practice 2A, 13e. American Petroleum Institute.

API. (1993). API RP 2A-LRFD Recommended Practice for Planning, design and constructing fixed offshore platforms. API Recommended Practice 2A, 20e. American Petroleum Institute.

API. (2000). API RP 2A Recommended Practice for Planning, design and constructing fixed offshore platforms. API Recommended Practice 2A, 21e. American Petroleum Institute.

API. (2011). API RP 2FPS Recommended Practice for Planning, Designing and Constructing Floating Production Systems. American Petroleum Institute.

API. (2014a). API RP 2A Recommended Practice for Planning, design and constructing fixed offshore platforms. API Recommended Practice 2A, 22e. American Petroleum Institute.

API. (2014b). API RP 2SIM Recommended Practice for Structural Integrity Management of Fixed Offshore Structures. American Petroleum Institute.

API. (2018). API RP 2MIM Mooring Integrity Management – Draft. American Petroleum Institute.

BSI. (2008). PAS55 Asset Management. British Standardisation Institute

Chakrabarty, J. (1987). *Theory of Plasticity*. New York: McGraw-Hill International.

Crisfield, M.A. (1996). *Non-linear Finite Element Analysis of Solids and Structures*. Chichester: Wiley.

CSWIP. (2018). Certification scheme for personnel compliance through competence. www.cswip.com (accessed 5 April 2018).

Department of Energy. (1977). First edition Guidance Notes for the Design and Construction of Offshore Structures. Department of Energy.

Department of Energy. (1984). Third edition Guidance Notes for the Design and Construction of Offshore Structures. Department of Energy.

Department of Energy. (1987). Department Of Energy United Kingdom Offshore Research Project – Phase II (UKOSRP II) Summary Report. HMSO, OTH-87-265.

DNVGL. (2016). DNVGL-RP-C208 Determination of structural capacity by non-linear finite element analysis methods. DNVGL, Høvik, Norway

EN. (2002). EN 1990:2002 Eurocode – Basis of structural design. European Standards, Brussels, Belgium.

EN. (2008). EN 473:2008 Qualification and certification of NDT personnel – General principles. European Standard

Ersdal, G. (2014). Safety of structures. Compendium at the University of Stavanger.

Ersdal, G. (2017). Safety barriers in structural and marine engineering. Invited paper for the symposium honoring Torgeir Moan. *Proceedings of OMAE 2017*. Trondheim, Norway.

EU. (2013). Directive 2013/30/EU on the European Parliament and of the Council of 12 June 2013 on safety of offshore oil and gas operations, European Union.

HSE (1990), HSE Guidance). *Offshore Installation: Guidance on Design, Construction and Certification*, 4e. London, UK: Health and Safety Executive (HSE).

HSE. (1996). The Offshore Installations and Wells (Design and Construction, etc.) Regulations 1996. Health and Safety Executive (HSE), London, UK.

HSE. (2005). HSE Offshore Installations (Safety Case) Regulation 2005, SI3117, Health and Safety Executive (HSE), London, UK.

HSE. (2007). Ageing semi-submersible installations – HSE Information sheet 5/2007, Health and Safety Executive (HSE), London, UK.

HSE. (2009). RR684 – Structural integrity management framework for fixed jacket structures, Health and Safety Executive (HSE), London, UK.

HSE. (2013). HSG 65 Managing for health and safety, Health and Safety Executive (HSE), London, UK.

HSE. (2015). HSE Offshore Installations (Offshore Safety Directive) (Safety Case etc.) Regulation 2015, SI398, Health and Safety Executive (HSE), London, UK.

IOGP. (2008). Asset Integrity – The Key to Managing Major Incident Risks. International Organisation of Oil & Gas Producers, London.

ISO. (1998). ISO 2394:1998 General principles on reliability for structures, International Standardisation Organisation.

ISO. (2006a). ISO 19903:2006 Petroleum and natural gas industries – Fixed concrete offshore structures, International Standardisation Organisation.

ISO. (2006b). ISO 19904:2006 Petroleum and natural gas industries – Floating offshore structures, International Standardisation Organisation.

ISO. (2007). ISO 19902 Petroleum and natural gas industries – Fixed steel offshore structures, International Standardisation Organisation.

ISO. (2013a). ISO/DIS 2394:2013 General principles on reliability for structures, International Standardisation Organisation.

ISO. (2013b). ISO 19900:2013 Petroleum and natural gas industries – General requirements for offshore structures, International Standardisation Organisation.

ISO. (2014). ISO 55001 Asset management – Management systems – Requirements, International Standardisation Organisation

ISO. (2016). ISO 19905:2016 Petroleum and natural gas industries – Site-specific assessment of mobile offshore units, International Standardisation Organisation

ISO. (2017). ISO/DIS 19901-9:2017 Structural Integrity Management, International Standardisation Organisation.

Moan, T. (1981). The Alexander L. Kielland accident. *Proceedings from the first Robert Bruce Wallace Lecture*, Massachusetts Institute of Technology, Cambridge, MA, USA.

Moan, T. (1983). Safety of offshore structures. In: *Proceedings of the 4th ICASP Conference*. Florence: Pitagora Editrice.

Moan, T. (1993). Reliability and Risk Analysis for Design and Operations Planning for Offshore Structures. Keynote lecture, ICOSSAR, Innsbruck, Austria (9–13 August 1993).

Moan, T. (1998). Target levels for reliability-based reassessment of offshore structures. In: *Proceedings of ICOSSAR*. A.A. Balkema.

NMA (2016). Regulations for Mobile Offshore Units. Haugesund, Norway: Norwegian Maritime Authority.

NOROG. (2017). NOROG GL 122 Norwegian Oil and Gas Recommended Guidelines for the Management of Life Extension, Stavanger, Norway

NPD (1977). *Acts, Regulations and Provisions for the Petroleum Activities*. Stavanger, Norway: Norwegian Petroleum Directorate.

NPD (1984). *Acts, Regulations and Provisions for the Petroleum Activities*. Stavanger, Norway: Norwegian Petroleum Directorate.

NPD (1992). *Acts, Regulations and Provisions for the Petroleum Activities*. Stavanger, Norway: Norwegian Petroleum Directorate.

NPD (2002). *Framework, Management, Facilities and Activities Regulations*. Stavanger, Norway: Norwegian Petroleum Directorate.

PMB Engineering (1993). *Hurricane Andrew – Effects on Offshore Platforms – Joint Industry Project*. San Fransisco, CA: PMB Engineering, Inc.

PSA (2004). *Framework, Management, Facilities and Activities Regulations*. Stavanger, Norway: Petroleum Safety Authority.

PSA (2016). *Framework, Management, Facilities and Activities Regulations*. Stavanger, Norway: Petroleum Safety Authority.

Sharp, J., Ersdal, G., and Galbraith, D. (2015). Meaningful and leading structural integrity KPIs. SPE Offshore Europe Conference, Aberdeen, Scotland (8–11 September 2015).

Skallerud, B. and Amdahl, J. (2002). *Nonlinear Analysis of Offshore Structures*. Baldock: Research Studies Press Ltd.

Søreide, T.H. (1981). *Ultimate Load Analysis of Marine Structures*. Trondheim, Norway: Tapir Forlag.

Søreide, T.H., Moan, T., Amdahl, J. and Taby, J (1982). Analysis of Ship/Platform Impacts. Third International Conference on the Behaviour of Offshore Structures, Massachusetts Institute of Technology, Boston, MA (2–5 August 1982).

Standard Norge. (2012). NORSOK N-001: Integrity of offshore structures. Rev. 8. Standard Norge, Lysaker, Norway.

Standard Norge. (2015). NORSOK N-006 Assessment of structural integrity for existing offshore load-bearing structures. 1e. Standard Norge, Lysaker, Norway

Standard Norge. (2017a). NORSOK N-003: Actions and action effects, 3e. Standard Norge, Lysaker, Norway.

Standard Norge. (2017b). NORSOK N-005 In-service integrity management of structures and maritime systems. 2e. Standard Norge, Lysaker, Norway.

3

Ageing Factors

Change is the law of life.

John F. Kennedy

Intelligence is the ability to adapt to change

Stephen Hawking

Nothing is so painful to the human mind as a great and sudden change.
Mary Wollstonecraft Shelley, Frankenstein

3.1 Introduction

Structures change from the day they are fabricated and these changes have to be managed in order to ensure that structures remain sufficiently safe. Some of these changes influence the structure and its safety directly. Examples of this may be fatigue, corrosion, material degradation, changes in loads and weight on the structure and how the structure is used.

In addition to the changes to the structures themselves, the loads and environment they operate in will change over time. Further, the way structures are used may change over the years which, as a result, will change the loading, the environment the structures are exposed to and possibly the configuration of the installation.

In addition, what is known about a structure will change, e.g. the type of information that has been retained from design and inspections of the structure. Further, the physical theories, mathematical modelling of these and engineering methods used to analyse the structures may change, typically as new phenomena are discovered.

Finally, the evaluation of offshore structures is also influenced by societal changes and technological developments, which may, for example, result in changes to the requirements that are set for offshore structures, taking into account obsolescence, lack of competence, and the availability of spare parts for old equipment.

These changes may be grouped into four different types:

- *Physical changes* to the structure and the system itself, their use and the environment they are exposed to (condition, configuration, loading and hazards).
- *Changes to structural information* for the structure and system (the gathering of more information about the structure and its state from inspections but also the loss of information from design, fabrication, installation and use).

Ageing and Life Extension of Offshore Structures: The Challenge of Managing Structural Integrity,
First Edition. Gerhard Ersdal, John V. Sharp, and Alexander Stacey.

- *Changes to knowledge and safety requirements* that modify the understanding of physics by models and methods used to analyse the structure and the required level of safety that the structure is expected to have.
- *Technological changes* that may lead to equipment and control systems used in the original structure being outdated, spare parts being unavailable and compatibility between existing and new equipment and systems being difficult.

These groups of changes are illustrated in Figure 3.1, where it is indicated that the physical and technological changes impact the safety and functionality of the structure directly, while structural information changes and changes to knowledge (engineering methods and physical theories) and safety requirements primarily change how the safety and functionality of the structure are understood. Further, it is indicated that physical and structural information changes apply to one specific structure, while technological changes and changes to knowledge and safety requirements are a result of societal and technological developments and are applicable to all structures.

Ageing factors may be sorted and illustrated in numerous ways. There is a need to differentiate first between those changes that directly modify the safety and functionality of the structure (upper two boxes in Figure 3.1) and those that primarily change the understanding of the safety of the structure (lower two boxes in Figure 3.1). While physical changes to the structure will have a direct impact on its safety, losing all the information about a structure (design reports, drawings, inspection results, etc.) does not actually change the safety of the structure but makes it difficult to document that the structure is sufficiently safe. Similarly, a new and improved understanding of the strength of a structure in the form of, for example, new formulae for joint capacity, only changes the understanding of the safety of the structure.

Technological solutions and equipment such as computers and control systems for managing ballast, mooring systems and stability will most likely continue to improve rapidly. The computers and control systems that were used when designing an old structure are in many cases outdated now. This will typically lead to a lack of replacement parts and may introduce compatibility issues if new components are required. There

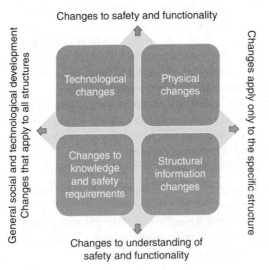

Figure 3.1 The four main elements of ageing of a structure.

are also technological developments in paint systems, corrosion protection systems and repair technologies that may improve the structural performance. All of these issues are referred to in this book as technological changes. Obsolescence is an aspect of these technological changes, when an object, service or practice is no longer fit for purpose even though it may still be in good working order. Typically, if an improved replacement is available, using the new technology in place of the old will directly impact the safety of the structure if not managed correctly.

The second type of differentiation of the ageing changes used in this book is between those that are applicable to one specific structure (right-hand two boxes in Figure 3.1) and those that are due to outside circumstances in society and hence apply to all structures (left-hand two boxes in Figure 3.1).

In other literature, similar types of changes are divided into physical and non-physical changes (Wintle 2010). This is a fully useful way of looking at these types of changes but gives slightly less framework for the identification of such changes. In other literature (ESREDA 2006), ageing changes are divided into: (i) degradation; (ii) obsolescence; and (iii) organisational. This is unfortunately the most widely used method to describe ageing changes but it lacks many important changes that should be included (such as other physical changes beyond degradation, changes to structural information and changes in knowledge and safety requirements). It is also unclear as to what is meant by organisational changes.

3.1.1 Physical Changes

The changes to a structure that are easiest to detect are visible changes such as corrosion, cracking, e.g. due to fatigue and creep, dents and damage due to accidental events. In general, the *physical changes* to the structure are the type of change that most people will think of first as structures get older.

Physical changes will also occur to the use, loading and environment that the structure is exposed to. Examples of this may be a changing corrosive environment inside a pipeline and changes to the loads and hazards to a structure. The addition of new modules and equipment may also be relatively easy to manage but a heavier inventory may be just as important a change and often more difficult to recognise. Slightly more difficult to detect may be changes due to settlement, tilt and induced forces due to differential settlement and subsidence. All these changes are *physical changes* as they lead to physical changes to the condition, configuration or loading on the structure. Any such physical changes may lead to the need for evaluating the structure again by a new analysis of loads, the strength and safety of the structure.

Physical changes and how to manage these is the core of this chapter and physical changes will be further discussed in Sections 3.2–3.7.

3.1.2 Structural Information Changes

Physical changes are by no means the sole challenge when existing structures are to be kept safe. The *information* about the actual structure, the loads on the structure, the hazards that they may be exposed to and the strength will also change. This does not modify the structure physically but it definitely changes the way of the safety of structures is perceived.

The knowledge of the design, fabrication, erection/installation and operation of a structure is vital in maintaining the integrity of the structure, e.g. in knowing:

- the loads for which the structure has been designed;
- the extent of marine growth for which the structure has been designed;
- which type of material has been used (strength, ductility, susceptibility to corrosion, and other degradation mechanisms, etc.);
- which parts of the structure have been inspected;
- where cracks and corrosion have been found;
- which members of the structure have been damaged;
- where the structure has been repaired;
- where repair welds were done in fabrication.

Over time the person(s) responsible for managing the integrity of the structure could have retired, been promoted or left the company. The archives of such information may have been lost or exist in a format that is no longer useful, etc. Hence, important information about the structure may be lost. In some cases, information about equipment may never have existed.

In the best examples, a well-kept database with information from design, fabrication, installation and use is available. More and more information about the structure and its performance will be gathered, typically from inspections, monitoring (e.g. accelerometers), updated load charts, updated metocean assumptions and practical experience from the use of the structure.

The reality is that often some or a significant amount of this data is missing or is lost. This results in a lower degree of the understanding of the safety of the structure and may make decisions on how to inspect, repair or modify the structure more difficult. The evaluation of whether the structure may be safe for further use will hence be difficult. The availability of good data about the structure and its materials provides increased confidence in the structure, its strength and hence its safety. Lack of such data will lead to uncertainty and reduced confidence in the safety of the structure.

The organisation responsible for the integrity of structures should be able to care for the competence, information and data needed to evaluate and document the safety of the structure or system. Necessary mitigating actions should be taken to cope with, for example, reorganisations, retirements and change of ownership, in order to ensure knowledge transfer. In addition, change of information storage methods should be addressed to ensure availability of information in the future.

3.1.3 Changes to Knowledge and Safety Requirements

Science and technology evolve and hence engineering methods for analysis and evaluation of structures also develop. Often these developments give improved methods for the analysis of structures that indicate an improved safety of the structure. However, new phenomena are also discovered that at times indicate that an existing structure is not sufficiently safe.

An example may be the likely assessments that bridge engineers had to perform on the bridges that they were responsible for in the aftermath of the failure of the Tacoma Bridge due to wind-induced vibrations. If in these assessments they found that their bridge was also susceptible to wind-induced vibrations, the safety of the bridge was not directly changed but the *understanding* of the safety of the bridge changed. Another example of such changes is the updated knowledge on wave heights and wave crest heights gained in the offshore industry, which can affect the safety of the structure.

Another interesting example of this type of ageing process relates to floating production, storage and offloading units (FPSOs). In 1998, FPSOs were a relatively new concept on the Norwegian Continental Shelf. New concepts often experience new phenomena on that were not expected during their design. A new phenomenon that was experienced with FPSOs was the so-called green water phenomenon, which is the over-running of the bow and side of the ship by waves that induces significant (and unexpected) wave loads on equipment on the deck of the FPSOs. This was unexpected for the designers of the first FPSOs, even though green water was a known phenomenon in merchant shipping. Measures implemented to reduce green water were typically so-called green water walls along the side of FPSOs but also operational restrictions on draft and trim were implemented (leading to requirements on maximum filling levels in oil tanks) in seasons of severe weather (mainly winter). These operational restrictions have been relatively successful. However, such restrictions have an impact on the economy and revenue and all such restrictions will always be under pressure and require a robust explanation by responsible competent persons both to justify these restrictions and their continued existence. Without such responsible competent person(s), such restrictions will disappear with time.

Society is in constant development and what was accepted as sufficiently safe many years ago may no longer be regarded as sufficiently safe. Often standards are updated in order to include findings from accidents and the identification of new phenomena. As societies develop they will normally become more safety aware and often more risk averse. A key aspect that changes the level of safety that is required is the experience of accidents. In the UK offshore industry, the regulation and requirements were dramatically changed after the Piper Alpha accident. A risk-based regulation was introduced and extensive risk analyses were required on a five-yearly basis, as part of a safety case. It is reasonable to claim that society did not accept the safety level that had been present in the offshore industry prior to the Piper Alpha accident.

These changes will not influence the safety of the structure directly but will clearly change how the safety of a structure is understood. They will also, to some extent, influence the decision on whether the structure is sufficiently safe.

3.1.4 Technological Changes

Developments in technology may lead to systems being obsolete, outdated, with limited spare part availability, compatibility problems with existing systems and lack of repairability. This type of ageing change is very relevant for marine systems in floating structures but less so for the structures themselves.

One of the longest running ship engines is believed to be the 35-year-old diesel engine on the MS *Lofoten* (NRK 2017). This engine has been in operation for close to 300 000 hours and is claimed to have travelled a distance similar to nine times to the moon and back. Intensive and systematic maintenance has been the key to keeping the engine running for so long. However, spare parts for the engine now have to be hand-made as spare parts are no longer available.

Similarly, a structure is designed, modelled, fabricated and erected/installed at a certain point in time, with the technology available at that point in time. In parallel with knowledge, technology will develop as a result of research and development

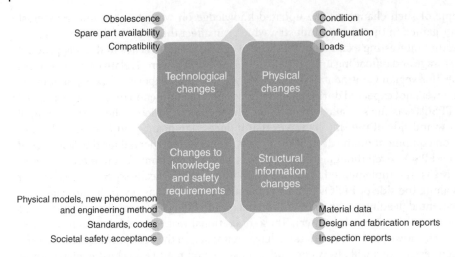

Figure 3.2 Ageing factors with examples.

and industry needs (e.g. new types of control systems for ballast and stability, etc.). Standards and regulations and hence designs will incorporate the best practice from this improved technology. As improved technology accumulates, the gap between new and original technology will increase. Eventually, this gap may become significant in a way that the original technology may be deemed unsafe (e.g. introduction of seat belts in cars). In addition, technology will develop and new types of equipment will be introduced. Production of spare parts for vintage versions of this type of equipment may then be stopped and hence it may be difficult or impossible to obtain them. Also, the compatibility between old equipment and newer developments may become difficult or impossible. Hence, technology based ageing may be defined as the effect of the structure being based on outdated technology.

Fortunately for the structural engineer assessing old structures, most of the early off-shore structures were built with a relatively good degree of conservatism. The extra material (mostly steel and concrete) is beneficial, and often crucial, when old structures are assessed for life extension. Also, the data, measurements, inspection results and other information that are available about the structures are very useful for the current evaluation of their safety.

Typical examples of the four types of changes used in this book are illustrated in Figure 3.2.

Physical changes are dealt with Sections 3.2–3.7, followed by a short overview of the non-physical changes in Section 3.8.

3.2 Overview of Physical Degradation Mechanisms in Materials

> *Degradation: the process of something being damaged or made worse*
> *Deterioration: the fact or process of becoming progressively worse*
> Oxford Learners Dictionary

Table 3.1 Degradation mechanisms in materials versus effect.

Effect	Metal loss/wall thinning	Cracking	Changes to material properties	Geometric changes
Degradation mechanism	Corrosion (chemical) –General –Pitting –Crevice –Corrosion under insulation –Galvanic –Stress corrosion cracking –Bacterial Flow induced metal loss (mechanical) –Erosion from solids Wear and tear	Fatigue Hydrogen related cracking –Blistering –HE –Stress corrosion cracking Creep	Hydrogen embrittlement Hardening –Overloads –Accumulated plastic deformation Environmental deterioration –Exposure period –Temperature –Bacterial	Dents from impacts Bowing (out of alignment) Permanent plastic deformations Corrosion

Degradation and deterioration have a common meaning but, in this book, degradation is used primarily as the process of something being damaged or made worse (*Oxford Learners Dictionary*). The degradation mechanism can result in metal loss (as uniform or localised attacks), cracking or changes to the material properties (e.g. increased brittleness). In addition, damage from impact, high loads from temperature expansion/contraction and quick pressure changes may influence the material to be more exposed to degradation mechanisms (e.g. dents from impact damage may form crack initiators that may lead to a fatigue crack) and can also lead to changes in the geometry of a structure or structural parts.

An absolute definition of which type of degradation the various deterioration mechanisms will cause is not possible. A general idea is given in Table 3.1.

Sections 3.3–3.5 describe metal loss and wall thinning due to corrosion, cracking due to fatigue, erosion, wear, and tear.

3.3 Material Degradation

3.3.1 Introduction

Materials selection at the design stage of offshore installations requires several factors to be taken into account, such as the ability to meet strength, temperature, fatigue and durability requirements of the structure. The properties of most materials used in offshore structures are now well understood as a result of operating performance and research. However, the fatigue properties of higher strength steels used, for example in the legs of jack-ups with yield strengths in excess of 600 MPa, are still less well known (HSE 2003). When older structures were designed, there were significant gaps in knowledge such as:

- through-thickness properties for large steel sections;
- the fatigue performance of large tubular joints and of concrete structures in seawater;
- long-term performance of reinforced concrete in deep water;
- grouting of very long prestressing ducts.

There are a number of deterioration mechanisms due to ageing; they principally are corrosion, fatigue, wear, and erosion, which are covered in this section.

3.3.2 Overview of Physical Degradation for Types of Steel Structures

The main degradation issues for jacket structures and for floating structures are given in Table 3.2 and Table 3.3, respectively.

NORSOK N-005 (Standard Norge 2017b) and the guidance document by Oil & Gas UK (O&GUK 2014) provide detailed information relevant for life extension and integrity management of ageing floating installations. The topics covered include:

- Hull structural integrity
- Hull watertight integrity
- Marine system integrity which includes ballast system, control system, cargo system, inert gas system and marine utilities (pumps, generators, etc.)
- Station keeping integrity

This information is a useful addition to this book and provides more detailed and highly relevant content for the life extension and integrity management of floating installations.

Table 3.2 Main degradation issues for jacket structures.

Jacket structure elements	Typical degradation specific for these elements
The main load-bearing members of the jacket structure	Fatigue is the most important degradation mechanism for the main load-bearing structure (as it has to withstand the cyclic loading, particularly from waves)
	Corrosion will be a typical problem for the substructure, usually involving cathodic protection and coating
	Ship collisions
	Dropped objects
Conductor guide frames	Fatigue damage is the most prominent degradation mechanism for submerged conductor guide frames, particularly taking into account out-of-plane loading
	For conductor guide frames in or above the splash zone, corrosion will be a major issue
Piles (foundation)	Fatigue in service and during pile driving
Pile sleeve connections	Fatigue
	Degradation of grout

Table 3.3 Main degradation issues in floating structures.

Floating structure elements	Typical degradation specific for these elements
Hull structural integrity	Fatigue is the most important issue for the main load-bearing structure as it has to withstand the cyclic loading, particularly from waves
	Corrosion will be a typical problem for the ballast and cargo tanks and external surfaces, usually involving cathodic protection and coating
	Ship collisions
	Dropped objects
Watertight integrity Doors, hatches, dampers, etc.	Wear and tear and corrosion
Marine system Ballast, control and cargo system, inert gas system, and marine utilities (pumps, generators, etc.)	Wear and tear and corrosion
Station keeping integrity	Fatigue
	Wear and tear
	Corrosion

An extensive study of physical degradation and inspection methods for mooring systems is given in HSE (2017). This is a comprehensive study of the important factors affecting the integrity of mooring chains.

The main degradation issues for topsides are given in Table 3.4.

3.3.3 Steel Degradation

3.3.3.1 Hardening Due to Plastic Deformation

When the overall stresses in a member or in an area exceed the elastic regime by more than localised plastic deformations, significant plastic deformations and plastic stress–strain behaviour in the structure will occur. Repeated plastic cycling can result in hardening or softening of the material. Cyclic hardening can result in increased resistance to static failure. Cyclic hardening results in a decreasing peak strain with increasing cycles whilst cyclic softening results in the strain range increasing and, ultimately, fracture.

Exposure to loading from the marine environment results in plastic cycling of the material in these plastic deformed areas. Plastic deformation under cyclic loading can cause the nucleation of fatigue cracks which can result in their initiation and propagation.

Many older platforms have over-utilised joints and members as a result of being designed to less rigorous criteria and with less knowledge about loading and strength behaviour than is available today. Also, increased topsides loading and changes in environmental loading over time will combine to cause over-utilisation of joints and members.

For further information on cyclic plastic behaviour, see DNVGL (2016b).

Table 3.4 Main degradation issues for topsides.

Topside elements	Typical degradation specific for these elements
The main load-bearing members of the deck structure (main-frame, deck-beam, primary structure of integrated decks, etc.)	Corrosion will be a typical problem for the whole topsides Fatigue is an important issue for the main load-bearing structure as it has to withstand the motions of the substructure and vibration from generators, process equipment, etc.
Flare booms	Wind will induce cyclic loading due to turbulence and vortex shedding. As a result, fatigue may be an important degradation mechanism In addition, the motion of the main structure will induce accelerations in the flare booms, which will also give cyclic fatigue loading
Helidecks	Similar to flare booms, wind will induce fatigue degradation The landing of helicopter may also induce cyclic loading
Derricks	Fatigue degradation is relevant for the same reasons as for flare booms. The rotational motion of the drilling will also induce loading on the derrick that may be cyclic In addition, the motion of the main structure will induce accelerations in the derrick, which will also give cyclic fatigue loading The main issue on derricks, however, is that derricks are normally constructed using bolted connections. These bolts have a tendency to loosen and fail under cyclic loading
Cranes	The main cyclic loading on cranes is due to the lifting operation, but also here wind may induce some cyclic loading
Support structures for riser, conductor, and caisson[a]	On floating structures this a major issue as the riser balcony has to transfer the motion of the main structure to the riser, inducing cyclic loading and potentially fatigue In fixed structures it is more the motion of the risers, conductors, and caissons themselves that need to be contained. This will also give cyclic fatigue loading
Load-bearing structures for modules	The motion and deformation of the deck will induce cyclic stresses into the load-bearing structure of the modules. Such stresses are most prominent for modules on ship-shaped platforms
Supports for safety critical items, such as temporary refuge, living quarters, etc.	Corrosion and fatigue similar to the main load-bearing part of the deck structure

a) This book does not include integrity management of the risers, conductors and caissons.

3.3.3.2 Hydrogen Embrittlement

Hydrogen embrittlement (HE), or hydrogen induced stress cracking (HISC), is the result of ingress of hydrogen atoms into the metal. When these hydrogen atoms recombine in voids in the metal matrix to form hydrogen molecules, they create pressure from inside the cavity and this pressure can reduce ductility and tensile strength up to a point where cracks open. In marine environments the principal sources of hydrogen in steel are from corrosion and cathodic protection (CP). Welding can also cause high hydrogen content if sufficient care is not taken during welding, for example by applying preheat and drying consumables. It has been shown that the uptake of hydrogen by steel in the marine environment is strongly influenced by the combined effects of CP and sulfate reducing bacteria (SRB) (Robinson and Kilgallon 1994; HSE 1998) CP produces hydrogen on the steel surface and its absorption is promoted by the biogenic sulfide produced by the SRB.

HE is a less common mode of failure in offshore structures than corrosion fatigue but it was found to be the cause of cracking that occurred in the leg chords and spud cans of jack-up drilling rigs operating on the UK Continental Shelf (UKCS) in the late 1980s (HSE 1991). This discovery of cracking caused considerable concern within the offshore industry and led to a significant research programme to understand the effects of hydrogen cracking, particularly where cathodic protection potentials are strongly negative, as can occur offshore. It was confirmed that hydrogen cracking can occur in high strength offshore structural steels due to HE under excessively negative cathodic potentials. This led to the Department of Energy (now the Health and Safety Executive) to provide guidance on limiting cathodic protection potentials for high strength steels offshore in vulnerable situations, the recommendation being that the CP level should be limited to a maximum negative value of −850 mV Ag/AgCl. This can be difficult to achieve in practice as outlined below. Steels for use offshore should be assessed for their vulnerability to hydrogen cracking and this guidance is contained in Section 33 of the HSE Guidance (HSE 1995).

High strength steels are being used increasingly in the construction of offshore platforms as they offer a significant weight saving compared with lower strength steels. At present, steels with yield strengths in the range 550–690 MPa are used for a variety of marine applications, particularly in the legs of jack-ups. As noted above, one disadvantage of using higher strength steels is the increased likelihood of HE, which can occur under static loads.

Susceptibility to HE is usually thought to increase with the strength of the steel and it is common practice to assess the likelihood of it occurring in a particular grade of steel on the basis of its strength or hardness. For a particular grade of steel, these parameters can give a useful first indication of its likely HE susceptibility. Quenched and tempered (Q&T) steels and controlled rolled (CR) steels can be produced with the same strength and hardness, yet they have very different microstructures and different susceptibilities to HE. However, it has been shown that when steels of different grades are considered together, the strength or hardness is often poorly correlated with HE susceptibility. It is concluded that HE susceptibility is more sensitive to the specific nature of the microstructure than to the strength level of the material.

There are a number of ways in which an actual strength level of a particular steel can be achieved. It is recommended that each steel should be considered individually and should be subjected to thorough testing before being accepted for use, particularly in critical locations and in circumstances that could lead to hydrogen charging. This is especially important in the case of steels for the construction of fixed jack-up platforms which are intended to remain in position for the life of the field and where inspection is more difficult to carry out. The effects of hydrogen charging from CP should be fully considered, particularly if there is a possibility that overprotection may occur. This can be mitigated by the use of potential limiting diodes or sacrificial anodes with a less protective potential. However, in practice both of these methods have limitations.

By controlling the composition and microstructure some modern offshore steels have been produced which have HE susceptibilities that are as low as those of BS4360 Grade 50D steels and significantly lower than would be expected on the basis of their higher strength alone.

The HE susceptibility of steels in seawater is frequently assessed using slow strain rate testing (SSRT) as it is a relatively rapid, comparative method. However, SSRT is a severe test (to failure) and does not accurately represent the conditions found offshore and it is important to appreciate the limitations of the technique as erroneous results can occur. For example, when welded specimens are tested the majority of the strain sometimes occurs preferentially in the softest part of the microstructure. This leads to failure in that region, whereas in practice HE usually occurs in the hardest region of the weld.

In terms of ageing structures with high strength steels, their history particularly with respect to cathodic overprotection needs to be taken into account to assess the possible degradation due to HE and any loss of strength and hence reduced integrity. The type of high strength steel and its known susceptibility to HE are also important factors in assessing the consequences from embrittlement due to hydrogen.

3.3.3.3 Erosion

Erosion can be defined as the physical removal of surface material due to numerous individual impacts of solid particles, liquid droplet or implosion of gas bubbles (cavitation). Erosion is a time dependent degradation mechanism, but can sometimes lead to very rapid failures. In its mildest form, erosive wear appears as a light polishing of the upstream surfaces, bends or other stream deflecting structures. In its worst form, considerable material loss can occur.

3.3.3.4 Wear and Tear

Wear has been observed to cause substantial degradation in mooring chains and pockets of fairleads. Significant loss of material can occur, sometimes in short periods of time (months), as a result of the rubbing of chain link surfaces. There is a subsequent loss of load-bearing capacity as the cross-sectional area of mooring chain links becomes too low to carry the applied load. Examples of wear are shown in Figure 3.3. This shows that mooring components are susceptible to fatigue cracks at various locations, including the flash weld and at or near the inter-grip area, as well as erosion and gouging.

Other failure mechanisms in mooring lines include corrosion, excessive tension from severe environmental loading, damage during operation, inspection and dropped objects, abrasion with the seabed and corrosion.

3.3.4 Concrete Degradation

3.3.4.1 Concrete Strength in Ageing Structures

In general terms, concrete increases in compressive strength with age as a result of hydration. However, concrete offshore structures introduce new issues; in particular, the long-term effect of deepwater immersion on strength. Initially when concrete is immersed in seawater at depth there will be an initial beneficial prestressing effect which reduces as the internal pore pressure develops. This pore pressure is developed in the concrete as it is a porous material and water can enter the pore structure. The time that this takes to develop is dependent on the permeability of the concrete which is generally

Figure 3.3 Wear and tear in chain links (unknown origin).

linked to strength. The presence of cracking also reduces the time for the internal pore pressure to be established. In terms of a typical offshore installation, the internal pore pressure will be fully developed in an ageing structure.

There are a limited number of results on the effect of deep water immersion on concrete strength. The UK Concrete in the Oceans programme exposed a number of cubes and cylinders in deep water (140 m) in Loch Linnhe in Scotland. The strength of these was compared with control cubes maintained for a similar time under shallow fresh water in a laboratory. After a short period of one-year immersion the two sets of results were similar but after 2.5 years of deep immersion the strength of the cubes was significantly different from the control cubes. The control cubes had either maintained or slightly enhanced their strength but the cubes exposed at depth showed a small loss in strength (compared with the control cubes). There was some concern about the testing regime as there were differences in time between testing the control cubes and deep-water cubes after removal from their location. The concern was that internal pore pressure in the deep-water samples had not been released before testing. However, thin slices from cylinders exposed at depth showed internal microcracking, particularly around the edge of coarse aggregate particles, which may explain the small loss in strength.

This loss of strength after deep water immersion has been confirmed by other tests (Haynes and Highberg 1979) which involved samples being tested after six years at depths of up to 1500 m. Typical strength loss was around ~10%. Other tests by Clayton (1986) confirmed a 10–15% loss in compressive strength of concrete after it had been pressurised to an equivalent depth of ~6000 m for a short period of only six days. These tests also showed an almost complete loss of tensile strength at this pressure, which may be of some concern in relation to cracking behaviour in deep-water concrete. However, no strength tests have been made on concrete under pressure.

Other work (Hove and Jakobsen 1998), which reviewed several test series on the effects of pore pressure, concluded that the effect was small and noted that many tests were undertaken using previously pressurised specimens but tested at atmospheric pressure.

As a result, some early offshore concrete structures were designed using a 10% reduction in strength. Later work concluded that neither compressive strength nor tensile strength was significantly influenced by pore pressure. However, the microscopic tests showing some internal cracking around aggregate particles is evidence that for those tests some deterioration of properties was evident. However, as noted above, these tests were undertaken at atmospheric pressure.

No test results are available for long term exposure at depth equivalent to an ageing structure. However, a 10–15% loss in compressive strength is unlikely to be important unless there are other issues such as the original concrete not meeting its design specification.

3.3.4.2 General

Concrete is naturally alkaline because of the presence of several hydroxides resulting from the reactions between the mix water and the Portland cement particles. This alkaline environment is important in providing protection against corrosion of the reinforcing steel (see Section 3.4.5). Loss of this alkalinity can occur through several ageing processes, which include ingress of chlorides and sulfate attack, which are discussed below.

Table 3.5 shows the correlation between the main parts of a concrete offshore structure and the primary degradation mechanisms applicable to these areas. Analyses have shown, however, that significant damage is required before significant loss of structural strength occurs to the legs of an offshore structure (Ocean Structures 2009).

Offshore concrete is generally very high quality (e.g. low water/cement ratio) and with thick covers to the outer layer of reinforcement (typically 70 mm in the splash zone,

Table 3.5 Deterioration mechanisms for parts of a concrete structure (Ocean Structures 2009).

Deterioration mechanism	Legs/towers/ shafts – general	Splash zone	Topsides	Steel concrete transition	Shaft/base junction	Storage cells	Foundation
Chemical deterioration	X	X			X	X	
Corrosion of steel reinforcement	X	X			X	X	
Corrosion of prestressing tendons	X	X			X	X	
Fatigue			X	X	X		
Ship impact	X	X					
Dropped objects						X	
Bacterial degradation	X	X			X	X	
Thermal effects					X	X	
Loss of pressure control					X	X	
Loss of air gap			X	X			
Scour and settlement							X

45 mm underwater) with limited permeability to seawater and post-tensioned to limit cracking. This permeability can be enhanced by the development of thin protective layers (brucite and aragonite – see below).

In some non-offshore concrete structures there has been evidence of alkali–aggregate reaction where the alkali nature of the cementitious material has led to a reaction with the aggregate used in the concrete mix. This has resulted in localised damage to the concrete and loss of integrity. Only certain aggregates lead to this type of damage and the crushed granite that was used for many offshore concrete structures has generally not been a problem. It is not known whether the aggregates used in all offshore concrete structures would be vulnerable to this type of damage.

Carbonation is a recognised ageing problem; this arises from the slow reaction between the CO_2 in the air and the hydroxyl ions in the concrete to produce carbonic acid. The alkalinity of the cover concrete is reduced by this acid, thus allowing further carbonation to take place at greater depths. Eventually this can lead to the embedded reinforcing steel losing the protection of the alkaline environment and to the possibility of corrosion. However, evidence from tests on bridge decks and other structures shows that with large depths of cover of high quality concrete this is unlikely to be a problem in the air zone of the concrete towers.

Sulfate attack is a reaction between the sulfates in seawater and calcium hydroxide in the hardened cement paste. The reaction products can be expansive leading to cracking and crazing of the concrete cover. This is a well recognised problem for marine structures and is usually designed out by, for example, including some pulverised fuel ash in the mix which reduces the permeability. However, this did not occur with the early offshore installations. Visual examination of the legs should detect any sulfate attack problems.

Tests in the 'Concrete in the Oceans Programme' (Department of Energy 1989) showed no significant chemical attack on concrete from deep water exposure after eight years (the length of the test programme).

Seawater contains a number of chemical ions which can participate in chemical reactions and which could lead to long-term degradation of the concrete. These include sodium, potassium and magnesium, as well as sulfate ions. However, concrete exposed to seawater can develop thin protective layers on its surface, which are mainly aragonite (calcium carbonate) and brucite (magnesium hydroxide). These layers protect the surface, modifying the permeability of the concrete and hence reducing the permeation of chlorides to the steel reinforcement.

3.3.4.3 Bacterial Induced Deterioration

Bacterial activity involving, for example, SRB in concrete structures containing both water and oil can lead to the production of acids, which attack the concrete. Significant loss of material has been shown in laboratory tests to occur when sufficiently acidic conditions exist (Department of Energy 1990).

This type of environment can exist in the concrete storage tanks, which are present in several concrete offshore structures. This is due to the presence of the oil–water mixtures from the operation of displacing the stored oil with seawater. SRB are known to grow rapidly under certain acidic conditions, which can cause loss of material reducing wall thickness. The rate of loss is dependent on the pH value increasing from pH 5 to more acidic levels (Department of Energy 1990).

Unfortunately, the storage tanks are very difficult to inspect due to very limited access and hence the level of damage from SRB is difficult to assess. The thick coating that is expected to exist on the inner walls of the tanks due to the presence of waxes in the oil may be a mitigating factor. Drill cuttings can accumulate around the lower sections of a concrete platform or on the roofs of the storage cells. These cuttings consist of oil and/or water based muds. Until the early 1980s, the muds were based on diesel oil, which was replaced in 1984 by low toxicity oil. As noted above, the presence of oil and water can encourage the development of bacteria, which are likely to be anaerobic, favouring the growth of SRB and possible deterioration of the concrete material.

3.3.4.4 Thermal Effects

The storage of hot oil in the concrete tanks at the base of many concrete installations can lead to thermal stresses that can produce cracking of the concrete. Concrete is vulnerable to significant temperature differences, which in this case arise from the hot oil on one side of the wall and cold seawater on the other. Tests have shown that temperature differences of up to 45 °C can be sustained with the correct design details (Department of Energy 1989). However, if the coolers fail (the oil is cooled before storage) or unusual conditions occur, oil with temperatures of up to 90 °C can be diverted into the storage cells, with potential cracking of the walls. Over a long period of operation these effects could accumulate. The increased stresses from thermal effects can lead to cracking of the concrete and overstress of steelwork in and around the walls and roofs of the storage cells, including the critical junction with the legs.

3.3.4.5 Erosion

Concrete subjected to an offshore environment can suffer erosion due to wave and wind loading or ice loading in certain offshore areas. The splash zone is particularly vulnerable due to the wave motion, temperature variations, etc.

The degree of erosion is very dependent on the quality of the concrete which is usually high for offshore structures (specified cement content, low water/cement ratio) with controlled permeability and the depth of cover (typically 75 mm in the splash zone). However, these specifications can only create a potential durability and the extent to which this is realised in practice is dependent on other factors, such as production control, curing, etc. Cracked sections or areas of low quality are more vulnerable to erosion. The main effect of local erosion of the concrete cover is earlier corrosion of the steel reinforcement with potential loss of strength (see Section 3.4.5).

Ice loading in certain offshore areas (e.g. Arctic structures) can lead to significant abrasion and needs consideration at the design phase. The phenomenon is caused by ice–concrete friction forces and results in gradual loss of concrete cover. In extreme cases, ice abrasion has brought about complete deterioration of the reinforcement cover of marine concrete structures. In some cases, special protection is required to minimise the effects of abrasion from ice sheets. Most offshore concrete structures operating in Arctic waters are not yet in the ageing phase but in due course ice abrasion will need to be addressed in life extension of such structures.

Damaged or eroded areas of concrete can be repaired using grout or specially formulated concrete.

3.4 Corrosion

3.4.1 General

Corrosion as a result of a chemical or electrochemical reaction between a metal and its environment leads to a deterioration of the material and sometimes its properties. The following basic conditions must be fulfilled for corrosion to occur:

- A metal surface is exposed to a potentially damaging environment (e.g. bare steel in physical contact with the local environment).
- Presence of a suitable electrolyte able to conduct an electrical current (e.g. seawater containing ions).
- An oxidant able to cause corrosion (e.g. oxygen, CO_2).

However, no corrosion will occur if one of these conditions is not present. Table 3.6 summarises prospective corrosion mechanisms for subsea oil and gas production equipment. More details of corrosion processes are given in DNV (2006).

3.4.2 External Corrosion

External corrosion of, for example, steelwork can occur in seawater, with the presence of absorbed oxygen leading to loss of material and reduced load carrying capacity. The rate of corrosion is dependent on the level of oxygen and the temperature of the seawater.

In the North Sea, the seawater is normally saturated with oxygen at approximately $6\,ml\,l^{-1}$. In other offshore areas, the oxygen level in deep water may be much less and corrosion would be inhibited.

External corrosion is usually mitigated by the use of a cathodic protection system and in some cases by the use of external corrosion coatings (see Chapter 5). The design of the CP system is dependent on the design life of the equipment and the type and quality of the external coating system in question. There are recommended levels of CP (e.g. −800 to −950 mV vs Ag/AgCl) (DNVGL 2015). On some installations there are more negative levels of protection and this overprotection can lead to the production of hydrogen with adverse effects on the steelwork. High strength steel (with yield strength greater than 500 MPa) is more susceptible to this overprotection and more stringent requirements are recommended for the level of CP (HSE 2003). Shielding can limit the effectiveness of

Table 3.6 Main corrosion mechanisms.

Corrosion mechanisms in an oil and gas environment	
Corrosion mechanism	Chemical reaction
O_2 corrosion	$2Fe + H_2O + 3/2\,O_2 = 2FeO(OH)$ (rust)
CO_2 corrosion	$Fe + H_2O + CO_2 = FeCO_3 + H_2$
Microbiologically induced corrosion (MIC)	$Fe + (\text{bacteria related oxidant}) \rightarrow Fe^{2+}$

Source: Based on DNV (2006).

the CP system, e.g. in areas where anode location is difficult. However, assuming the CP protection is effective there should be limited loss of material due to external corrosion. However, the CP system in not effective in the splash zone and alternative means of protection are required (e.g. coatings, plus a corrosion allowance). A number of coating systems have been used offshore with epoxy-based systems being more typical.

Some components (e.g. chains) are not usually provided with a CP system due to the complexity of the chain unit and a corrosion allowance is normally provided in design. This allowance will depend on the expected corrosion conditions and the expected life. In some areas where they may be a limited supply of oxygen (e.g. contained areas) the corrosion rate may be lower and a smaller corrosion allowance necessary.

3.4.3 Various Forms of Corrosion

3.4.3.1 CO_2 Corrosion

Carbon steel, can be subjected to CO_2 corrosion. The corrosion rate is dependent on the partial pressure of CO_2, the temperature, the flow regime and the water in-situ pH. The corrosion is a time dependent degradation mechanism and generally is localised in the form of pitting. CO_2 corrosion can be managed by the use of corrosion inhibitors and/or by pH stabilisation of the process fluid, which is primarily applicable for pipelines.

3.4.3.2 Environmental Cracking Due to H_2S

Environmental cracking due to the presence of H_2S, arising from either bacterial activity or the presence of drill cuttings, is linked to sulfide stress cracking (SSC) and carbon steel is susceptible to SSC. The likelihood of SSC is dependent on a number of factors, including the partial pressure of the H_2S, the total tensile stress, chloride ion concentration and the presence of another oxidant. SSC is not expected to occur below a critical partial pressure of H_2S. However, for partial pressures above this limit there is an increasing likelihood for SSC and the environmental condition is termed 'sour'. Cracking is the resulting failure mode and can be abrupt in nature. SSC is controlled by specification of the material properties (particularly hardness) and the manufacturing process. For susceptible materials, environmental cracking is more likely to occur during the initial phase of production. However, older installations may experience a souring of the wells (the produced amount of H_2S increases) and the production environment turns from sweet to sour. This can lead to a higher probability for environmental cracking which is dependent on the material properties and the changed service conditions.

3.4.3.3 Microbiologically Induced Corrosion

Microbiologically induced corrosion (MIC) is a form of degradation that can occur as a result of the metabolic activities of bacteria in the environment. The bacteria that cause MIC can accelerate the corrosion process because the conditions that apply already have elements of a corrosion cell. SRB are the most aggressive microorganisms that enhance the corrosion of steel. SRB bacteria live in oxygen-free environments, making use of sulfate ions in the seawater as a source of oxygen. H_2S is produced as a waste product from the SRB, producing a local corrosive environment in connection with the bacteria. MIC has been observed on steel (e.g. anchor chains on the seabed) buried within seabed sediments. The likelihood of MIC occurring is difficult to predict as it depends on the availability of nutrients, water temperature and local flow conditions.

3.4.4 Special Issues Related to Corrosion in Hulls and Ballast Tanks

The integrity of floating installations depends on intact hulls and ballast tanks; corrosion is the main threat to this. Ballast tanks are particularly vulnerable as seawater is used for ballast purposes. Corrosion protection systems are usually provided to limit the extent of corrosion, either by coatings or CP or by a combination of both. Surveys required, for example, for the maintenance of Class, require thickness measurements of critical parts of the structure and monitoring of the protection system. DNVGL-RU-OU-0300 (DNVGL 2018) lists the relevant survey requirements and has a special section on 'Special provisions for Ageing Units' which is mainly concerned with fatigue and the introduction of the 'fatigue utilisation index' (FUI). When the FUI exceeds one then special measures are required including additional surveys. In addition, systematic thickness measurements need to be performed at renewal surveys and inspection of the corrosion protection system to establish its effectiveness. For critical areas, a detection system is recommended to establish any water ingress as a result of corrosion or cracking. DNVGL-RP-B101 (DNVGL 2015) for corrosion protection for floating production and storage units lists both design and survey requirements for corrosion protection systems. This document points out that it is a challenge to provide more than 10 years' service life for the corrosion protection of an FPSO. Whilst more traditional vessels dock every five years for detailed inspection and repair, an FPSO will be in continuous operation for its service life. The DNVGL (2015) document points out that it is therefore necessary to develop an improved specification for the corrosion protection of an FPSO with a service life of 10 years or longer. This should be based on experience for the corrosion protection of fixed offshore structures with design lives exceeding 25 years. Clearly, a case for life extension would need evidence of thickness measurements of critical areas, as well as evidence of the continuing performance of relevant CP systems and coatings protecting against corrosion.

Ballast tanks are significantly exposed to corrosion, especially in areas where the use of anodes cannot provide the required protection. Oil tanks are also exposed to corrosion, especially if the oil has a low pH value. This has a tendency to form pitting corrosion in the bottom of a tank. If inert gas produced from oil is used (with potential sulfide content), corrosion may also be a problem for the deck head. Hence, corrosion protection in the form of coating and anodes is important in such areas.

3.4.5 Concrete Structures

3.4.5.1 Corrosion of Steel Reinforcement

In concrete offshore structures the continuing integrity of the steel reinforcement is an essential requirement. Steel which is embedded in concrete should normally be protected from corrosion for long periods, provided there is a good depth of high quality cover over the steel. The recommended level of cover for the steel reinforcement is 70 mm in the splash zone and 45 mm in the underwater zone. As noted in Section 3.3.4, concrete is a permeable material and hence the chlorides in seawater will penetrate to the steel reinforcement in the longer term. Activation of the reinforcement with loss of passivity can occur when sufficient chlorides reach the steel surface and if sufficient oxygen is available this will usually lead to corrosion. This is typically the case for the splash and air zones where, over a period of time, corrosion will progress leading to expansive products which usually cause spalling of the concrete cover. Following spalling of the cover, corrosion usually occurs more rapidly unless repaired. This type of corrosion is very typical of many marine structures.

The splash zone is particularly vulnerable with a plentiful supply of both oxygen and seawater. In this location corrosion is more likely to occur.

Cracking of the concrete cover is a process which allows easer ingress of seawater to the embedded steel. Typically, the basis of the design process requires control of cracking after construction, with a limit of 0.1 mm for the splash and atmospheric zones and 0.3 mm for the submerged zone. Additionally, cracking of the cover can occur during operation when tensile stresses are present due to a number of factors including:

- high static stresses from storms;
- fatigue, loss of localised reinforcement;
- external damage (dropped objects or ship collision);
- the presence of expansive corrosion products from corrosion of the reinforcement.

As noted earlier, this can be followed by spalling, where localised sections of the concrete are lost, leading to further water ingress. Evidence of significant cracking from visual inspections indicates the potential for degradation of the structure and the need for local repair.

Underwater there is limited availability of oxygen and this limits the degree of corrosion and, as shown by laboratory work (Department of Energy 1989), the corrosion products are non-expansive and do not usually lead to spalling of the cover. This has its disadvantages, however, in terms of detecting the corrosion using visual inspection. Intense localised corrosion has been observed in underwater concrete test sections:

- where there has been a local breakdown of passivity (e.g. from cracking);
- where there is low concrete resistivity (e.g. from long term immersion in seawater or de-icing salts);
- where electrical conductivity through the steel reinforcing network is present which can act as a link between the anode and cathode;
- an efficient cathode is also required (which may be in the splash zone where there is sufficient oxygen to support the high local corrosion rate).

Similar localised intense corrosion of the reinforcement has also been seen in bridge decks which have become saturated with chlorides from de-icing salts.

Where cyclic stresses are present in the structure, fatigue of the steel reinforcement can occur leading eventually to localised loss of reinforcement and potential cracking of the concrete cover. This can be exacerbated by the presence of seawater. Cracking of the cover can lead to further damage to the steel reinforcement due to corrosion.

Accidental damage such as ship collision and dropped objects can cause localised damage to the towers (ship collision) or to the tops of the caissons (dropped objects), resulting in seawater ingress and local corrosion of the steel reinforcement. It is important therefore to detect such damage soon after it occurs and repair the section to minimise further damage from corrosion.

In most offshore concrete structures there is a CP system, which is present to protect the attached steelwork to the concrete columns. The steel reinforcement is usually connected to this, despite early attempts at the construction stage to isolate the reinforcement. This connection can happen particularly with unintended electrical connectivity to flowlines and pipelines as well as other external attachments. This has led to a higher than planned drain on the sacrificial anodes for some of the early structures and in some cases these have had to be replaced. In later structures the reinforcement was deliberately connected to the attached steelwork to benefit from CP.

CP has an important advantage in that it protects the reinforcement to some extent, minimising the level of corrosion where seawater has permeated to the steel or where cracking is present. Current design criteria for CP systems for offshore concrete structures recommend or require a minimum allowance of 1 mA m^{-2} for the reinforcing steel. Maintenance of the CP system is therefore a basic requirement to minimise the corrosion reaction in concrete structures. However, the CP system has minimal protection for the splash zone which is the most vulnerable to corrosion. This is the reason for higher than normal levels of concrete cover in this zone. The seawater reacts with the concrete to form both aragonite (calcium carbonate) and brucite (magnesium hydroxide), which tend to build up at the mouth of the cracks and hence can limit the ingress of seawater with beneficial effects.

3.4.5.2 Corrosion of Prestressing Tendons

High strength prestressing tendons are required to maintain the structural integrity of the concrete structure, particularly in the towers. These tendons are placed in steel ducts which are grouted following tensioning. The degree to which grouting has been effective, given the long ducts and in some case their horizontal orientation, has led to concerns that seawater can penetrate into the ducts and cause corrosion of the very high strength tendons with potentially serious local loss of prestress. A review of the durability of prestressing components (HSE 1997) concluded that the first tranche of concrete offshore structures (pre-1978) was more vulnerable to corrosion of the prestressing tendons as later platforms benefited from improved grouting materials and procedures. It was also considered that there would need to be significant loss of prestress ($\sim$40%) in a leg before it could fail under typical design wave loading. These failures would also need to be in the same section area to be a danger. In land based structures, failures have tended to occur near anchorages or construction joints.

3.5 Fatigue

3.5.1 Introduction

Fatigue is characterised by cumulative material damage caused by numerous loading cycles during the service life, resulting in crack initiation and propagation. Fatigue cracks will occur from discontinuities and defects in areas with high stress. Welded joints with high stress concentrations are a typical example. Fatigue failure is normally considered to occur when a through-thickness crack forms.

Fatigue cracking is a time-dependent and accumulative degradation mechanism and hence cracks normally should be expected to occur late in the life of a structure. However, there is evidence that cracking can occur within the design life. Evidence of this has been seen when defects from the fabrication process remain. An example of this is the fatigue crack that formed in the *Alexander L. Kielland* flotel and eventually caused the capsize of this platform in 1980.

Fatigue failure is a significant hazard to offshore structures subjected to cyclic loading (e.g. wind and wave loading) in harsh environmental conditions. An early review of

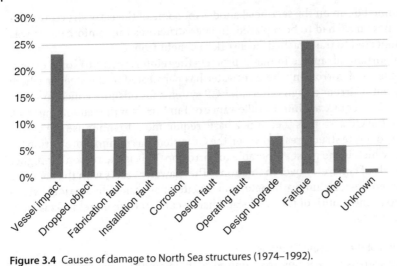

Figure 3.4 Causes of damage to North Sea structures (1974–1992).

repairs (MTD 1994) of offshore structures in the UK sector of the North Sea showed that fatigue was the main single cause of damage requiring repair for this early dataset, as illustrated in Figure 3.4. Another review (MTD 1992) of repairs to North Sea structures highlighted 40 incidents where significant repairs had been required as a result of fatigue costing many millions of pounds.

A number of incidents of severed members and fatigue failures in installations operated in the North Sea have occurred over the many years of operation and they have been a matter of some concern (MTD 1994). The importance of fatigue from data obtained in the period is shown in Figure 3.4.

A study into the fatigue damage in offshore structures was performed by Stacey and Sharp (1997) based on the HSE database of major structural damage to fixed structures in the UK sector of the North Sea. This study included 170 incidents, including significant fatigue failures, in the period 1972–1991, based on a review of 174 platforms. Some of this damage was not repaired and would therefore not have been included in the Marine Technology Directorate (MTD) database mentioned above. The data included severances, known and suspected through-thickness cracks, dents deeper than 50 mm and bows with a maximum deflection greater than 100 mm. The HSE database contains information indicating major fatigue cracking in 20 of the 174 fixed steel platforms installed in the Northern, Central and Southern North Sea in the period 1966–1984. The structures were located in water depths of 22–186 m. The data shows that the majority of damage was detected during the first 10 years after installation, indicating under-design for fatigue or poor fabrication. Further, this study indicates that for early structures fatigue damage was detected throughout the platform life, even as early as in the first five years, whereas for structures installed after 1980 there had been less damage. In addition, this study (Stacey and Sharp 1997) indicated that it was evident that there was a steady increase in the number of failures with accumulated service life though the data confirmed that there was a particularly high incidence of fatigue failure in the early years.

The primary methods for assessing fatigue life are the *S–N* approach and the fracture mechanics approach. These fatigue assessment methods have developed significantly over the last 40–50 years. The *S–N* method in particular is empirical and based on laboratory tests to develop characteristic design curves for the fatigue assessment. The design curves include a safety margin to allow for the inherent uncertainty in the test data. Hence, the design curve is normally derived by the logarithmic mean curve minus two standard deviations, providing a probability of failure of 2.3% based on the test data.

A large proportion of the *S–N* tests were performed under constant amplitude loading, while actual components experience a range of different stress amplitudes due to wave action. The models and methods used to evaluate the fatigue stress ranges are also empirical and introduce additional uncertainty into the fatigue assessment.

Reliable fatigue assessment procedures are required if the likelihood of fatigue failure is to be evaluated. The fatigue assessment should be used to enable the implementation of appropriate control measures. Fatigue safety within the required design life is regarded to be achieved by:

- Designing structural components with fatigue lives meeting the planned life and allowing for the required fatigue safety factor (design fatigue factor, DFF).
- Fabricating these structures with a minimum of discontinuities and defects.
- Having the ability to inspect when and where necessary.
- Having the ability to repair developed fatigue cracks that could affect overall structural integrity.

Detailed information on the structural condition is an important requirement in the structural integrity assessment of ageing installations and life extension. During the operation of offshore structures, inspection is carried out to identify any damage (e.g. cracking). Failure to detect fatigue damage has resulted in major structural failures, the *Alexander L. Kielland* accident mentioned above being a particularly prominent example. This led to a major effort to develop suitable fatigue design and assessment methods in the 1980s and 1990s and this has resulted in a substantial reduction in the amount of fatigue damage being found. Thus, fatigue failure is an important consideration throughout the lifecycle of the structure, i.e. during design, fabrication and the service life, and hence in the integrity management of ageing structures.

Early fatigue designs were based on less stringent criteria than what is available now, resulting in premature cracking, as shown in Figure 3.5, which required in many cases expensive repairs and in a few cases in serious accidents. As an example, no allowance for the environmental effect of seawater and blanket[1] stress concentration factors (SCFs) were normally used in the 1970s. At that time, offshore structures were designed on the basis of the requirements for those in the Gulf of Mexico, where fatigue damage is less of a problem.

The application of more modern fatigue assessment methods to ageing existing structures has led, in some cases, to reduced calculated fatigue lives compared with what was initially believed to be the case. However, in many of these cases there is no evidence of early fatigue cracking as predicted by the improved methods. This emphasises that the methodology of fatigue analysis is not intended to predict a precise fatigue life but rather it provides a tool intended to ensure that the likelihood of cracks in the design life is reduced to an acceptable level. The uncertainties in fatigue analysis and how these are dealt with in the standardised fatigue analysis methods is further described in Section 4.5.

---

1 Blanket stress concentration factors are non-specific and not relating to the geometry of the detail.

Figure 3.5 Large fatigue crack in welded tubular joint. Source: J.V. Sharp.

A considerable amount of work has been published on the fatigue of marine structures. The most relevant background references are Gurney (1979), Maddox (1991), Almar-Næss (1985), Lotsberg (2016), and DNVGL-RP-C203 (DNVGL 2016a).

3.5.2 Factors Influencing Fatigue

The primary factors influencing fatigue are:

- Defects and discontinuities in the material.
- The presence of cyclic stresses.
- The environment the material is operating in.

Fatigue damage has been shown to occur as a result of fabrication defects being present, particularly at welds and where there are stress concentrations such as at geometrical discontinuities. Defects are inherent to the welding process and thus the crack initiation stage may be shorter in welded connections compared with that in non-welded components.

Stress concentrations in welded offshore components can result in stresses that can be many times greater than the nominal stress and lead to cracking at these locations. Stress concentrations are often analytically described by SCFs. High SCFs typically occur at transitions, joints, connections, built-in discontinuities (e.g. thickness changes), supports, etc. Background information on stress concentrations is given in DNVGL-RP-C203 (DNVGL 2016a).

Cracking in offshore steel jacket structures has occurred mainly at welded nodal joints and at circumferential butt welds (Figure 3.5). The magnitude of the SCF in these areas can vary according to the type of connection and the type of fatigue loading. Parametric formulae enable SCFs to be calculated for joint type and loading for many types of connections.

The environment the material is used in also has a significant effect on the fatigue life. Most early data on fatigue resulted from testing in air and it was recognised subsequently that immersion in seawater had a detrimental effect on fatigue performance. Testing showed that the fatigue life could be reduced by a factor of at least two unless CP or induced current (IC) were present. For lower stress ranges, CP is more beneficial: the fatigue life is found to be very close to the life found in dry air. This approach now forms the current guidance for the fatigue design of welded connections.

Another important effect on fatigue capacity is the so-called thickness effect. Increasing the size of a given type of fatigue specimen while maintaining all other parameters will in general cause a decrease in fatigue strength (Morgan 1983). This thickness effect has been quantified as a result of testing large specimens and is now included in the design requirements.

It is also important to notice that fatigue damage is very dependent on the cyclic stress range, and fatigue damage is proportional to the cyclic stress range to the third power.

3.5.3   Implications of Fatigue Damage

The primary implication of fatigue damage is reduced structural strength, increased fatigue crack growth, an increased chance of brittle or ductile fracture and that water ingress can occur to members. As an example, a crack in a tubular joint will reduce the static strength and remaining fatigue life (Mohaupt et al. 1987). For example, a through-thickness crack can reduce the static strength by 40% (Stacey et al. 1996). The consequences of fatigue failure need to be properly understood in the management of ageing installations.

Through-thickness cracking can be followed by member severance and loss of stiffness in the local structure. This will lead to load redistribution which will cause other components to be more heavily loaded, possibly resulting in more rapid fatigue cracking in other areas. Hence, multiple cracking may occur and, depending on the level of redundancy, ultimately the structure may fail. Thus, as both fatigue life predictions and component strength are affected by load redistribution (Noordhoek et al. 1987), it is important that due consideration is given in the development of the structural integrity management plan to the possibility of total member failure occurring after penetration of the wall and to its consequences.

It is apparent that the influence of load redistribution on fatigue life can lead to unexpected failures as the fatigue of intact structures does not account for this load redistribution after fatigue failure. It is also possible that fatigue cracks may initiate and propagate at fabrication defects which are not necessarily located in those joints identified from the structural analysis as being critical. A lack of information on such defects will give an incorrect view of the structural integrity of the installation. This places additional emphasis on the need for an understanding of the system performance.

Member failure may not occur only due to a fatigue crack but also as a result of reduced capacity due to a fatigue crack being exposed to a high level of loading, e.g. due to wave loading or ship impact. These loads could also lead to local collapse of greater consequence in areas with significant amounts of fatigue cracking. It should be noted that this occurrence of multiple cracking, as might occur towards the end of life of a jacket structure, and its impact on structural integrity is not normally considered in the integrity management of ageing installations. This is an omission that could have significant consequences.

Fatigue is not restricted to the substructure of a jacket or the hull of a floating structure. A particular example is cracking in conductor guide framing in early structures as a result of under-design resulting from a lack of understanding of the importance of out-of-plane wave loading in regions of high stress concentration. The problem has been overcome in newer structures with better design practice.

In addition to fixed structures, semi-submersible structures are sensitive to fatigue cracking. The following areas are normally regarded as very important in this respect (Standard Norge 2017b):

- End connections of braces (to columns, pontoons and the deck).
- Connections between individual braces.
- Individual butt welds and transitions in braces:
  - conical to tubular sections;
  - square to tubular sections;
  - between tubular and square sections.
- Connection between pontoons and columns, including:
  - transition between pontoons and pontoon nodes;
  - transition between ring pontoons;
  - connections between pontoons and columns;
  - transition between square to tubular sections;
  - cast or forged transitions.
- Connection between columns and upper hull, including:
  - transition between square to tubular sections;
  - cast or forged transition pieces.
- Discontinuities such as terminations of sponsons, blisters, etc.
- Internal support structure, stiffeners, brackets, thickness transitions and notches.
- External brackets.
- Supports for anchor line fairleads, chain stoppers and winches (anchor windlasses).
- Topside support connections to internal structure in the hull (including crane pedestals, flares, helideck and drilling derrick).

Important structural details in ship-shaped structures that are vulnerable to fatigue are structural connections with high stress concentrations located in areas with localised high dynamic pressures. As an example, parts of the side shell are fatigue sensitive. Relevant details where cracking has been found are given in NORSOK N-005 (Standard Norge 2017b), including:

- Highly stressed structures such as moonpool corners and openings and termination of related stiffeners.
- Turret support structure.
- Transverse web frame terminations including related bracket toes.
- Toe and heel of horizontal stringers in the way of transverse bulkheads.
- Upper and lower hopper knuckles.
- Other relevant discontinuities and terminations in bulkheads, stringers and frames.
- Knuckle lines in side shell.

- Bilge keel supports and terminations.
- Portions of bulkheads and frames subjected to concentrated loads.
- Hull envelope openings, penetrations and attachments.
- Hull envelope butt welds/block erection welds and corresponding scallops.
- Longitudinal stiffener support connections to frames and bulkheads (especially at side shells).
- Stiffened plates in the side shell.
- Cross-tie supports/terminations and corresponding brackets.
- Cut outs for doors, pipe and cable penetrations, air ducts, etc..
- Discontinuities of longitudinal stiffeners in bottom plating and at bilge radius.
- Topside support connections to main deck and related internal structure in the hull (including crane pedestals, flares and helideck).

Other areas sensitive to fatigue include topside components, such as deck legs, deck trusses and girders; these are important to the overall integrity of the structure and consequence of failure. With ageing, fatigue can become an issue and inspection of topside components vulnerable to fatigue needs to be part of a maintenance programme. Inspection of these components is easier as special methods required for underwater use are not needed.

3.5.4 Fatigue Issues with High Strength Steels

Fatigue test data and guidance has mainly been developed from tests on medium grade structural steels. Higher strength steels are also used, particularly in jack-ups, with yield strengths up to 700–800 MPa. Much less test data are available for these materials and there is evidence that the effect of seawater on the fatigue performance of these high strength steels is more detrimental, with a greater susceptibility to hydrogen cracking (see Section 3.3.3.2).

The fatigue behaviour of high strength offshore structural steels, defined generally as those with a yield strength greater than 425–450 MPa, has been the subject of several studies and published work from Europe, USA and Japan has been reviewed (Standard Norge 2015). Fatigue crack propagation data for a range of high strength offshore structural steels have been compared with data for conventional BS 4360 50D structural steels. Information for parent plate and heat affected zones and the influence of environment, i.e. air, free corrosion and CP potentials ranging from −800 to −1100 mV has been considered. The data for parent materials indicate that comparable or slightly improved performance compared with that of structural Grade 50D steels can be obtained in both air and seawater environments suggesting that the ageing performance would be similar. However, the combined detrimental influence of high mean stress and overprotection levels (i.e. −1100 mV) can cause crack growth rates to be increased significantly compared with those observed in air and the reduction of fracture toughness due to hydrogen embrittlement (HE).

The results suggest that an essential element of maintaining the integrity of high strength steel structures is the maintenance of the corrosion protection system to achieve acceptable corrosion fatigue and crack growth performance. To achieve this, some codes and standards require protection potentials to be controlled in a narrow band (−770 to −830 mV Ag/AgCl), which is difficult to achieve in practice. However, it should be noted that the data reviewed above are for relatively modern high strength steels and therefore may not necessarily be representative of all high strength steels used offshore – it would be expected that, in some cases, older steels used in early structures may have inferior performance to modern steels, which needs to be taken into account in assessment procedures for ageing.

3.5.5 Fatigue Research

It was recognised in the 1970s that design to resist fatigue was a necessary requirement for offshore structures in various geographical regions around the world, particularly in northern European waters. This was highlighted by some catastrophic failures such as the loss of the *Alexander L. Kielland* semi-submersible in 1980. It was recognised that the fatigue of welded components, particularly large welded joints with both simple and complex geometries found in offshore structures, was not well understood and large research programmes were conducted to provide new data and enable the development of fatigue guidance which was first introduced in the early 1980s. The importance of fatigue is highlighted by the amount of research that has been performed over the last 40 years (e.g. HSE 1987, 1988; Noordhoek and de Black 1987; Dover and Glinka 1988) and in the effort that has been made to develop new standards and guidance. Substantial programmes of research on the fatigue performance of welded tubular joints, comprising the generation of *S–N* curves in air and seawater environments, the development of SCF equations for simple and complex tubular joint configurations and the development of fracture mechanics methods have been undertaken.

As the environment in the North West Atlantic is the most fatigue prone area, much of the fatigue research was conducted in the UK and Norway. The most notable research programme was the United Kingdom Offshore Steels Research Project (UKOSRP), the first phase of which commenced in 1975 and led to significant amendments on fatigue design in the Department of Energy Guidance Notes, which at that time were the basic requirements that designers and certifiers needed to comply with for North Sea structures (HSE 1988). This programme, funded mainly by government, led to significant improvements in knowledge on fatigue. The main structural materials tested were grade 50 steels which were the most widely used offshore. A main development from this programme was the introduction of an *S–N* curve for welded tubular joints (the T curve) – previous *S–N* curves were based on the testing of plates and did not properly reflect the features and performance of tubular joints with hot spot stresses.

A Norwegian five-year project was initiated in 1977 for intensified research on fatigue of offshore steel structures. This project was funded by the Norwegian Research Council. The aim of the project was to establish more reliable methods for the fatigue analysis of steel structures and developing an understanding of mechanisms and conditions governing the fatigue life of steel structures. Many types of offshore steel structures were included in the project, including semi-submersibles, tension leg platforms and jack-ups, in addition to jacket structures. The resulting *Fatigue Handbook* (Almar-Næss 1985) formed the basis for fatigue design of offshore structures in Norwegian waters (and possibly elsewhere) for decades.

A second phase of UKOSRP research into the fatigue of steels in a seawater environment was carried out which investigated further the effect of seawater on fatigue, the effect of thickness on fatigue performance and other factors (HSE 1987). The results from this programme led to a new edition of the Department of Energy Guidance Notes (HSE 1995). This was recognised as the most advanced design guidance on fatigue and was adopted by the Norwegians in NORSOK N-004 and later by the

International Standards Organization (ISO) in the preparation of the standard on design and operation of offshore fixed structures, ISO 19902 (ISO 2007).

Later, the FPSO Fatigue Capacity joint industry project (2001–2003) addressed the fatigue capacity of FPSOs. Numerical investigations and fatigue testing were performed in order to improve the accuracy, robustness and efficiency of finite element modelling and hot spot stress evaluation for typical FPSO details (Lotsberg et al. 2001; Lotsberg 2004; Lotsberg and Landet 2004).

Research on the fatigue performance of offshore structures has included work that is of generic applicability but also specifically relevant to ageing installations. Key areas include the use of weld improvement techniques to extend the fatigue life and the development of inspection methods to detect the onset of fatigue cracking.

3.6 Load Changes

Another physical change influencing the safety of the structure directly are changes in loads on the structure. Modifications and addition of modules and more equipment are typical examples of such load changes. However, the wave and wind loading on structures also change and need to be addressed in structural integrity management and especially in life extension assessments. The wind loading will generally change due to the change in wind area, e.g. as modules are added. Wave loads will change as a result of increased marine growth with time and due to the addition of structural parts, risers, conductors and caissons in the wave affected zone. If a structure experiences subsidence (typically due to reservoir contraction), the waves may expose new previously unexposed areas of the structure and the overturning moment of the loading will increase. This is most critical if the topside of the installation is impacted by waves. Wave in deck will dramatically increase the wave loading on the structure.

Changes to wave loading have also increased due to new knowledge on wave height statistics, updated understanding of wave kinematics and new knowledge about slamming pressures from waves.

Global warming may affect the wave and wind climate and hence affect the loading on the structures. At present, there are insufficient data to provide information on the effect on wind and wave climate as a result of global warming. NORSOK N-003 'Actions and action effects' (Standard Norge 2017a) indicates that the effect of global warming should be accounted for in structures that are planned to be operated for more than 50 years.

3.6.1 Marine Growth

The loading on a structure from waves and current is significantly influenced by marine growth and fatigue assessments, in particular, will have to be updated with updated marine growth data (Birades and Verney 2018). As per the Morison equation given in ISO 19902 (ISO 2007), the hydrodynamic force F on a member of diameter $\varnothing$ is the sum of the drag force F_d and the inertia force F_i:

$$F = F_d + F_i = C_d \cdot \frac{1}{2} \rho_w U \cdot |U| \cdot \phi + C_m \cdot \rho_w \pi \frac{\phi^2}{4} \cdot \frac{\partial U}{\partial t}$$

Hence, the drag force is proportional to $\varnothing$ and the inertia force is proportional to $\varnothing^2$.

Depending on geographical area, the thickness of marine growth can be significant and have a major impact on fatigue life due to the increased wave and current loads.

3.6.2 Subsidence and Wave in Deck

In 1969, Phillips Petroleum, now ConocoPhillips, discovered the Ekofisk field in the Norwegian sector of the North Sea. Several facilities were installed on the Ekofisk field during the 1970s. By 1984 it was discovered that platforms in the Ekofisk field had subsided several metres (subsidence is motion of the seabed as it sinks to a lower level relative to a datum such as sea level, e.g. due to the extraction of oil and gas in the reservoir beneath). This change had many implications for the safety of the structures supporting the facilities:

- Loads were acting at different places of the structures, possibly also with the potential of hitting the deck of the facilities.
- Fatigue damage may have become more severe in some areas, especially on horizontal conductor frames.
- Corrosion was primarily occurring in the splash zone (and to a lesser extent above water) for structures with CP. New areas of the structure were as a result of the subsidence exposed to the splash zone.

Most of the facilities on the Ekofisk field were jacked up late in the 1980s, relieving the operator of some of the concerns mentioned above. However, the field has continued to subside and work has been ongoing for many years on installations and their structures in order to achieve sufficient safety.

3.7 Dents, Damages, and Other Geometrical Changes

Structures can accumulate damage during service, mainly from ship collisions, dropped objects or extreme weather. Impact by dropped objects or swinging loads during lifts by cranes and similar devices also constitutes an important hazard scenario for an offshore installation. This damage is in the form of dents, or bows and sometimes these are associated with cracks. Figure 3.6 shows a typical dented member. These dents have

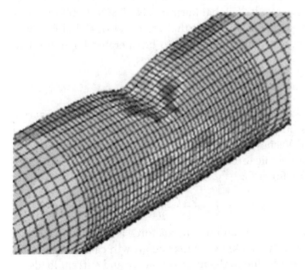

Figure 3.6 Illustration of dented member.

a significant effect on member buckling capacity and on the static capacity of tubular joints, beams and stiffened plates.

Periodic inspection will identify some of this damage, some of which will be repaired, but experience indicates that a significant part of the damage will remain either undiscovered or unrepaired. During the installation's lifetime multiple damage sites could build up to an extent where collectively they may weaken a structural member.

Surveys of data on dents and bows have shown, as indicated in HSE (1999), that dents found during inspection can be up to 300 mm in depth but more typically dents are in the range between 10 mm and 60 mm in depth. Some of these are associated with bows which can be quite large, up to 500 mm in magnitude. Some more serious dents and bows are associated with cracking and dependent on the local stress magnitudes which can lead to fatigue cracking which requires specific monitoring and possible repair.

The structural effect of dents and bows on strength has been investigated both through physical tests and by modelling. ISO 19902 (ISO 2007) contains a section on the effect of dents on tubular members. Equations are given for the effect on strength and stability for dented members subjected independently to axial tension, axial compression, bending or shear. More details of these are given in Chapter 4.

HSE (1999) identified threshold levels of dent and bow damage as follows:

- Dent damage of 12.5 mm when an incident that is likely to cause damage is known to have taken place.
- Dent damage of 38 mm in the absence of an alert of an incident. Such damage could be detected by a general visual inspection (GVI) survey provided that marine growth has not occurred to obscure the dent.
- Bow damage of 130 mm following an immediate response to a known incident.
- Bow damage of 350 mm when a routine inspection is being carried out.

For the determined thresholds for dents, typical strength reductions are up to 6% and 17% for legs and braces, respectively. Corresponding values for bow thresholds are as high as 37% and 68%, respectively (HSE 1999). These indicate that structural integrity could be compromised in the cases where the damage is not found following an incident and a major storm occurs. The report concludes that for dents and bows current North Sea inspection practice is not necessarily effective in detecting such damage. There is also published evidence that in practice many dents and bows are detected during inspections following an unreported accident. An early review (MTD 1994) showed that concerning vessel and dropped object damage only 17% was detected during routine inspection or by chance, rather than as a response to an event. A key aim of periodic inspection is to ensure that any degradation stays within acceptable limits.

For ship-shaped structures, longitudinals in side shells may be bent or twisted due to a local impact resulting in a local reduced structural capacity. In semi-submersibles, the columns are most vulnerable to local denting. A bent and twisted ring stiffener or girder will reduce the global buckling capacity of the column.

In ageing structures, accidental damage can accumulate, and the combined effect of multiple dents and bows can reduce the resistance of the structure significantly. This combined effect is not covered in current codes and standards. Hence, this places pressure on operators to detect individual incidents of damage during regular inspections.

Inspection for accidental damage includes both the initial detection and a requirement to characterise the damage. The former is likely to be by GVI or close visual inspection

(CVI) using remote operated vehicles (ROVs) or divers (see Appendix A). Photographs can establish the general shape of the damage but photogrammetry is needed to provide more detail. Characterisation involves measurement of the dent depth and its extent and the degree of bowing. It is also important to establish the presence of any cracks. Measurement techniques for dents and bows include use of taut wires, straight edges, ovality and profile gauges. Accurate dimensions of the damage are needed both to establish the degree of loss of strength and also to plan repair techniques (if required). The most important task in the evaluation of dents and bows is to identify those that have an impact on strength locally or globally.

The repair of dents and bows is addressed in Section 5.4.4. Typical techniques include grout filling of members and grouted clamps for large dents. Bows are more difficult to repair, unless by member replacement.

3.8 Non-physical Ageing Changes

The non-physical ageing changes are as indicated in Section 3.1:

- Technological changes (obsolescence, spare part availability and compatibility issues).
- Changes to knowledge and safety requirements (physical models, new understanding of phenomena and changed societal safety acceptance).
- Structural information changes (changes in availability of material data, design, fabrication and inspection reports).

3.8.1 Technological Changes (Obsolescence)

Technological changes have already been mentioned in Section 1.2.1. In general, this type of change is a result of the general technological development in society. It will materialise in equipment and systems (e.g. control systems) becoming outdated. This is also often called obsolescence.

Oil and Gas UK (2012) defines obsolescence as 'structures, systems or components passing out of usefulness as a result of changes in knowledge, standards, technology or needs'. Obsolescence is typically characterised by the absence of necessary spares and technical support in the supply chain. This can also occur due to changes in standards or technology but excludes physical deterioration. In reality, for equipment related to the production of oil and gas, typically this can result from any of the following (or a combination of more than one):

- vendors will no longer support equipment;
- vendors are out of business;
- spare parts are no longer available;
- upgrades made to software systems;
- equipment functionality no longer meets industry requirements or standards.

Offshore operators need to be aware of current and potential obsolescence areas affecting the asset so that suitable plans can be developed to deal with issues, minimising any unexpected problems.

O&GUK (2012) indicates the following acceptable actions:

- Replacement with suitable alternatives (this may involve sourcing from non-standard routes, e.g. from other equipment in the asset/company portfolio).
- Replacement with new component (functionality may be different from an existing component and any changes require to be assessed).
- Identification of an alternative solution to replacement (note that identification of an alternative solution can reduce the consequence of failure to an acceptable level in some instances and all such proposals must be fully risk-assessed).
- Identify operating mode which allows equipment to be used for the remaining the required life.

Where like for like replacement is not possible, it is important to carry out a thorough review applying management of change principles to ensure that safety implications are properly understood and considered. Interfaces between new and original equipment should also be carefully evaluated, considering the facility lifecycle and potential for any future replacement of the original equipment.

For many structures, this type of change will have limited impact, as they do not include computers and other equipment that can be obsolete. However, floating structures, for example, rely on ballasting systems with computers, watertight doors and hatches, pumps, and vents that clearly could experience technological changes and obsolescence. In many offshore concrete structures, the stresses in the base storage units and the adjacent parts of the leg are affected by the internal pressure in the cells. This pressure is controlled by the water level in the ballast header tank inside one of the legs. There is normally an underpressure in the cells compared with that in the surrounding seawater. Loss of this could arise from a failure in the ballast water system or oil storage pipework which could lead to possible overstress of the structure under severe storm conditions. In addition, failure of the ballast water system can lead to loss of underpressure.

3.8.2 Structural Information Changes

Availability and use of structural and material data from the design, fabrication and operation of the structure are important in managing structures. If these data are available and in use they will be of great benefit for the understanding of the structure and will provide the necessary information for good decisions on how to manage the structure. If these data are partly or totally missing, the safety of the structure will be understood to a lesser degree and this will affect decisions on how to inspect, repair, or modify the structure.

Good structural information data give increased confidence in the structure, its strength and hence its safety. They also provide important value in analysis and assessments of the structural safety (e.g. for life extension). Lack of such data will lead to uncertainty and reduced confidence in the safety of the structure.

Structural information data disappear for several different reasons. A lot of information, especially on older structures is to a large extent in the memory of individuals. If data are archived, archives may have been lost or exist in a format that is no longer useful. Changes in ownership and contractors as well as the UK change from certification to verification in the period 1996–1998 have contributed to a loss of corporate knowledge.

This typically includes loss of data on the design criteria, the history of inspection and repair (including accidental damage). It is important, therefore, that operators ensure that continuity of knowledge and experience is maintained. An important consideration for life extension is adequate knowledge of the installation, both of its current state and of the original design criteria (e.g. the design, fabrication and installation resumé).

3.8.3 Knowledge and Safety Requirement Changes

Knowledge changes are the changes that occur in the understanding of structural models and methods. The experience of structural engineering associated with offshore installations has developed over the several decades since such installations were placed in the sea and continuing research and development has increased understanding of both loading and structural performance. Subsequently, knowledge has increased substantially with the consequence that the criteria used in the original design of older structures has changed. In life extension of ageing structures, such changes need to be taken into account and the newest knowledge (and hence often the newest standards) need to be used in an assessment of an ageing structure for life extension.

Indeed, the HSE Guide on the Management of Ageing of Older Installations makes the point that in an operator's review of a safety case (the so called thorough review) 'new knowledge and understanding, e.g. awareness of the risk highlighted by industry or HSE safety; recognition and inclusion of findings from relevant research' need to be considered (HSE 2009). Particular examples given in the HSE report include an improved understanding of system performance following single and multiple member failure and the effects on fatigue life due to load redistribution. Similar requirements also exist in other regulatory regimes, such as in Norway.

Over the several decades that offshore installations have been installed many technological developments have taken place, for example in materials understanding and performance, in fabrication techniques and in inspection and maintenance. For example, steel quality has improved, particularly in through-thickness properties and weld techniques are better established. New codes and standards have evolved which reflect these changes. These differences need to be taken into account in an assessment of an ageing structure. The HSE guidance referred to above states that in assessing a safety case of an ageing structure a comparison should be made of the effect of using older codes and standards, comparing the case against current standards and HSE guidance and industry practice for new installations. As a result, the effect of any deficiencies should be evaluated such that any reasonably practicable improvements to enhance safety can be made and implemented. Eventually, this gap may become significant in a way that the original design may be deemed unsafe.

Another important area where knowledge development has improved is in risk management, which in the early days of offshore installations design was basic and lacked the level of understanding that has evolved from experience and research. A major factor in this area is an understanding of fire and explosion and their potential effect on the installation. There is now a better understanding and techniques available for explosion modelling, leading to higher predicted overpressures than was known during the design of many early installations.

Bibliographic Notes

Section 3.2 is based on Hörnlund et al. (2011). Section 3.4 is partly based on DNV Ageing Material, a report produced for the Petroleum Safety Authority (PSA) Norway (DNV 2006). Section 3.3.3.2 is based on HSE (2003).

References

Almar-Næss, A. (ed.) (1985). *Fatigue Handbook – Offshore Steel Structures*. Trondheim, Norway: Tapir, Norwegian Institute of Technology.

Birades, M. and Verney, L. (2018). Fatigue analysis, lifetime extension and inspection plans. In: *Proceedings of the 3rd Offshore Reliabilty Conference (OSRC2016)*, Stavanger, Norway (14–16 September 2016). Trondheim, Norway: Norges teknisk-naturvitenskapelige universitet, Institutt for marin teknikk (IMT).

Clayton, N. (1986). Concrete strength loss from water pressurisation. Conference on Concrete in the Marine Environment, Concrete Society.

Department of Energy (1989). Concrete in the Oceans Programme – Co-ordinating Report on the Whole Programme, Concrete in the Oceans Technical report no. 25, UEG/CIRIA. HMSO, London.

Department of Energy (1990). Concrete Offshore in the Nineties – COIN, Offshore Technology report OTH 90 320. HMSO, London.

DNV (2006). Report No. 2006-3496 Material risk – ageing offshore installations. DNV, Høvik, Norway.

DNVGL (2015). DNVGL-RP-B101 Corrosion protection of floating production and storage units. DNVGL, Høvik, Norway.

DNVGL (2016a). DNVGL-RP-C203 Fatigue design of offshore steel structures. DNVGL, Høvik, Norway.

DNVGL (2016b). DNVGL-RP-C208 Determination of structural capacity by non-linear finite element analysis methods. DNVGL, Høvik, Norway.

DNVGL (2018). DNVGL-RU-OU-0300 Fleet in service. DNVGL, Høvik, Norway.

Dover, W.D. and Glinka, G. (eds.) (1988). *Fatigue of Offshore Structures*. Warley, UK: Engineering Materials Advisory Services Ltd.

ESREDA (2006). *Ageing of Components and Systems*. European Safety Reliability and Data Association (ESREDA).

Gurney, T.R. (1979). *Fatigue of Welded Structures*. Cambridge, UK: Cambridge University Press.

Haynes, H.H. and Highberg, R.S. (1979). Long-term deep-ocean test of concrete spherical structures – results after 6 years. Technical report no. R 869. Civil Engineering Laboratory, Port Hueneme, CA.

Hörnlund, E., Ersdal, G., Hinderaker, R.H., et al. (2011). Material issues in ageing and life extension. *Proceedings of the ASME 2011 30th International Conference on Ocean, Offshore and Arctic Engineering, OMAE 2011*, Rotterdam, the Netherlands (19–24 June 2011).

Hove, K. and Jakobsen, B. (1998). Pressure effects on design of deep water concrete platforms. *Proceeding of the Second International Conference on Concrete under Severe Conditions, CONSEC'98*, vol. 3. E & FN Spon.

HSE (1987). OTH 87 265 United Kingdom Offshore Steels Research Project – Phase II: Final Summary Report. HMSO, London.

HSE (1988). OTH 88 282 The United Kingdom Offshore Steels Research Project – Phase 1: Final Report. HMSO, London.

HSE (1991). OTH 91 351 Hydrogen Cracking of Legs and Spudcans on Jack-Up Drilling Rigs – A Summary of Results of an Investigation. HMSO, London.

HSE (1995). *HSE Guidance Offshore Installation: Guidance on Design, Construction and Certification*, 4, third amendment, 1995e. London: Health and Safety Executive (HSE).

HSE (1997). OTO 97 053 The Durability of Prestressing Components in Offshore Concrete Structures. HSE Information Service.

HSE (1998). OTH 98 555 A Review of the Effects of Sulphate Reducing Bacteria in the Marine Environment on the Corrosion Fatigue and HE of High Strength Steels. HMSO, London.

HSE (1999). OTO 1999 084 Detection of Damage to Underwater Tubulars. Health and Safety Executive (HSE), London.

HSE (2003). Review of the Performance of High Strength Steel Used Offshore. Health and Safety Executive (HSE), London.

HSE (2009). Information Sheet Guidance on Management of Ageing and Thorough Reviews of Ageing Installations, Offshore Information Sheet No. 4/2009. Health and Safety Executive (HSE), London.

HSE (2017). RR 1090 Mooring Integrity for Floating Offshore Installations Joint Industry Project. Phase 2: Summary. Health and Safety Executive (HSE), London.

ISO (2007) ISO 19902, *Petroleum and natural gas industries – Fixed steel offshore structures*. International Standardisation Organisation.

Lotsberg, I. (2004). Recommended methodology for analysis of structural stress for fatigue assessment of plated structures. OMAE-FPSO'04–0013, International Conference, Houston, TX.

Lotsberg, I. (2016). *Fatigue Design of Marine Structures*. New York: Cambridge University Press.

Lotsberg, I. and Landet, E. (2004). Fatigue capacity of side longitudinals in floating structures. OMAE-FPSO'04–0015, International Conference, Houston, TX.

Lotsberg, I., Askheim, D.Ø., Haavi, T., et al. (2001). Full scale fatigue testing of side longitudinals in FPSOs. *Proceedings of the 11th ISOPE*, Stavanger, Norway.

Maddox, S.J. (1991). *Fatigue Strength of Welded Structures*. Abingdon, UK: Cambridge Universty Press.

Mohaupt, U.H., Burns, D.J., Kalbfleisch, J.G. et al. (1987). Fatigue crack development, thickness and corrosion effects in welded plate to plate joints, Paper TS3. In: *Proceedings of the Third International ECSC Offshore Conference*, Delft, The Netherlands (15–18 June 1987). Elsevier.

Morgan, H.G. (1983). The Effect of Section Thickness on the Fatigue Performance of Simple Welded Joints, Springfields Nuclear Power Development Laboratories Report No. NDR941(S).

MTD (1992). Probability-based fatigue inspection planning, Report 92/100. Marine Technology Directorate (MTD), London.

MTD (1994). Review of Repairs to Offshore Structures and Pipelines. Report 94/102. Marine Technology Directorate (MTD), London.

Noordhoek, C. and de Black, J. (1987). *Proceedings of the Third International ECSC Offshore Conference*, Delft, The Netherlands (15–18 June 1987). Elsevier.

Noordhoek, C., van Delft, D.R.V., and Verheul, A. (1987). The influence of the thickness on the fatigue behaviour of welded plates up to 160 mm with attachment or butt weld, Paper TS4. In: *Proceedings of the Third International ECSC Offshore Conference*, Delft, The Netherlands (15–18 June 1987). Elsevier.

NRK (2017). Hurtigrute-motor er trolig verdens lengstgående (in Norwegian). www.nrk.no (accessed 23 June 2017).

O&GUK (2012). *Guidance on the Management of Ageing and Life Extension for UKCS Oil and Gas Installations*, issue 1, April 2012. London, UK: Oil & Gas UK.

O&GUK (2014). *Guidance on the Management of Ageing and Life Extension for UKCS Floating Production Installations*. London, UK: Oil & Gas UK.

Ocean Structures (2009). OSL-804-R04 Ageing of Offshore Concrete Structures. Ocean Structures.

Robinson, M.J. and Kilgallon, P.J. (1994). HE of cathodically protected HSLA steels in the presence of sulphate reducing bacteria. *Corrosion* 50 (8): 620–635.

Stacey, A. and Sharp, J.V. (1997). Fatigue damage in offshore structures – causes, detection and repair. *Proceedings of the 8th International Conference on the Behaviour of Offshore Structures, BOSS 1997*, Delft, The Netherlands.

Stacey, A., Sharp, J.V., and Nichols, N.W. (1996). Static strength assessment of cracked tubular joints. In: *Proceedings of the 15th International Conference on Offshore Mechanics and Arctic Engineering*, Florence, Italy. New York: American Society of Mechanical Engineers.

Standard Norge (2013). NORSOK N-004 Design of steel structures, Rev. 3 February 2013. Standard Norge, Lysaker, Norway.

Standard Norge (2015). NORSOK N-006 Assessment of structural integrity for existing offshore load-bearing structures, 1e. Standard Norge, Lysaker, Norway.

Standard Norge (2017a). NORSOK N-003: Actions and action effects, 3e. Standard Norge, Lysaker, Norway.

Standard Norge (2017b). NORSOK N-005 In-service integrity management of structures and maritime systems, 2e. Standard Norge, Lysaker, Norway.

Wintle, J. (2010). The Management of Ageing Assets, Presentation given at TWI Technology Awareness Day, TWI, Cambridge, UK (14 October 2010).

4

Assessment of Ageing and Life Extension

4.1 Introduction

The purpose of the assessment of an existing structure is to ensure that the structure is acceptable for further use, particularly for life extension, taking into account changes that have occurred and other factors that may undermine confidence in its integrity. Assessment of an existing structure is typically needed when:

1. Changes have occurred to the condition of the structure, such as:
   - Deterioration due to time-dependent processes such as corrosion and fatigue.
   - Structural damage by accidental loads or an extreme weather event.
2. Changes have occurred or are planned to the loading on the structure, such as:
   - Increased loading from updated metocean data, the addition of new modules and loading areas, an increased number of risers or conductors, increased wind loading areas and wave in deck loads as a result of subsidence.
3. Changes have been made to the use of the structure, such as:
   - Increased service life.
   - Changes to the structure to accommodate modifications in its use (e.g. manning levels and operation).
   - Increased size of supply vessels.
   - Exceedance of original design life.
4. Changes have been made to the requirements to the structure, such as:
   - Requirements for increased safety (increased importance to owner, public or society).
   - Changes that have been implemented in standards and regulations (e.g. due to new knowledge about structural failures).
5. When there is doubt about whether the assumptions underlying its original design are fulfilled, such as:
   - The structure has not been inspected for an extended period of time.
   - Unexpected degradation has been observed.
   - The structure has been subject to accidental or otherwise unforeseen extreme loads (e.g. weather events).
   - Similar structures have shown unsatisfactory performance.
   - New knowledge and revised design codes.

In order to ensure sufficient safety in extended use of the structure, a method of evaluating the structure has to be established. The primary way of evaluating the safety of

Ageing and Life Extension of Offshore Structures: The Challenge of Managing Structural Integrity,
First Edition. Gerhard Ersdal, John V. Sharp, and Alexander Stacey.
© 2019 John Wiley & Sons Ltd. Published 2019 by John Wiley & Sons Ltd.

a structure in use is by performing design code checks (partial safety factor method for different limit states) using current standards and taking into account inspection and survey results. Other recognised ways of evaluating existing structures may include:

- non-linear ultimate capacity checks;
- comparison with other structures;
- proof-loading (not easily applicable for offshore jacket structures and other types of sub-structures).

Typically, assessment and analysis should include evaluation of the effect of changed requirements for the use of the structure, validation of the design assumptions and assessing the effect of possible deviations from these on the structural performance, as well as assessment of the condition and residual capacity and service life of the structure. Any part of the structure that is not meeting the assessment requirements may be improved (e.g. ground or peened), strengthened (e.g. grouted or reinforced) or replaced with new structural parts.

Other ways of reducing the risk to personnel from structural failure are to introduce risk prevention and mitigation procedures, e.g. evacuation procedures. Evacuation may be an option if the main hazard is caused by a predictable event such as excessive wave loading or wave in deck loading.

The requirement of an assessment is to document that the structure is sufficiently safe for further use. A reasonable goal, used under some regulations such as the NOR-SOK N-series of standards (Standard Norge 2015), is that the level of safety required is the same as a newly designed structure. Other regulations such as API RP 2A (API 2014) allow a somewhat decreased safety level for some ageing structures, often with some limitations in e.g. manning and use. This difference is reflected in the historical development of regulations, and the attitude of different societies concerning safety.

Certain information should be available for the assessment of an existing structure, including knowing:

- The structure and marine systems (correct drawings) in order to do the right calculations.
- The degradation history and prediction of future degradation.
- What is inspectable, replaceable and what is not.
- Up to date regulations and standards for the assessment of ageing structures for life extension.
- Any relevant developments in technology.
- How to perform analysis for the assessment of ageing structures for life extension (which differs from the design of new structures).

4.1.1 Assessment Versus Design Analysis

There are several differences between a *structure being designed* and *an existing structure subject to an assessment*. The first difference is the available information about the structure, which includes previous performance data and the current condition of the structure. This information can be taken into account as a possible method for evaluating the continuing safety of the structure.

The second difference is that the costs associated with improving an existing structure will normally be much greater than improving a structure still at the design stage.

In the design of new structures, the cost of adding a little extra steel or concrete in the structure is limited and does not necessarily justify expensive advanced engineering analysis. This typically means that it is acceptable to use standard design methods. These include methods such as linear elastic analysis, standardised code checks of most members, nodes and details, and stress concentration factors (SCFs) will often be taken from simplified standardised formulae. However, in assessing an existing structure, the cost of any modifications to the structure are relatively large. Hence, more advanced structural analysis methods are often preferable. This typically requires that members, nodes and details are modelled in detail in advanced finite element analysis programs and the use of non-linear structural analysis[1] and structural reliability analysis (SRA).[2]

New structures are designed according to design codes that take into account the assumed uncertainties by statistical characteristics of loads and strength and partial safety factors. In contrast, an existing structure can be measured, inspected, tested, instrumented and sometimes proof-loaded. In theory, all information relevant for assessing the condition and performance of an existing structure can be collected. However, in practice proof-loading is in most cases unrealistic and the gathering of a large amount of data is costly and requires significant effort. In addition, the fact that an existing structure has survived a number of years subjected to given loads and exposures contains information of value in the assessment situation.

In order to ensure sufficient safety in extended use of a structure, taking into account the above information, a method for evaluating the structure has to be established. A number of different procedures have evolved, which are described in the next section.

4.2 Assessment Procedures

4.2.1 Introduction

The purpose of assessing an existing structure should be to ensure that the integrity of the structure is still acceptable, even though it may be in a degraded state. In order to achieve this, several assessment procedures are proposed in standards and guidelines (e.g. ISO 19902, NORSOK N-006). Figure 4.1 indicates the main steps typically found in such assessment processes (as previously shown and explained in Section 1.5).

The obvious part is to assess the 'as is condition' (including deterioration such as corrosion, cracks and dents) and, based on this assessment, existing drawings and computer models should be updated in order to perform the necessary analyses. For these analyses to be realistic, the load description has to be updated taking into account any changes in loads or load specification (e.g. due to updated environmental criteria, weight

---

1 Most design is performed by the use of linear analysis, which assumes small displacements and linear elastic material behaviour. In contrast, non-linear analysis takes into account large deformations, plasticity and work hardening. Non-linear analysis requires much more computational effort and competence of the user, as the input to the analysis and the interpretation of the results is much more difficult. In addition, the superposition principle is not applicable when using non-linear analysis. See Section 4.4.4 for more details.
2 The standard design methods are so-called semi-probabilistic, where the uncertainty in for example, loads and strength are modelled by statistically defined characteristic values. In a structural reliability analysis all these uncertainties are modelled by their probability distribution. Hence, this can be called a probabilistic approach to structural design. See Section 4.7 for more details.

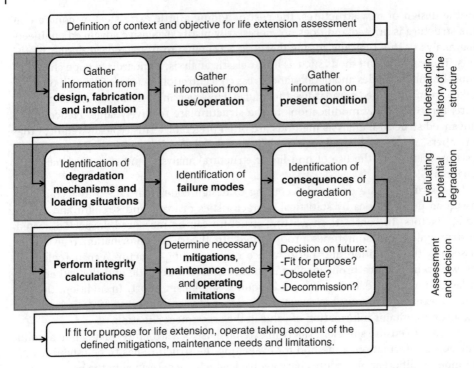

Figure 4.1 Life extension assessment process.

increases and subsidence). The engineering methods for calculating loads and strength of the structure may also have changed since the original design and the effects of these changes also need to be assessed.

If the degradation history (such as number of cracks and extent of corrosion) of the structure and its components is available, this may be used to indicate trends in the degradation process. For example, if degradation has developed slowly over time or has increased rapidly in the recent past. Also, the history of incidents and accidents should also be evaluated, along with their influence on such aspects as the strength of the structure. Any positive performance history (such as the absence of cracks during inspections) will also be important information for reducing uncertainty about the structure. Without such information on degradation and performance history, uncertainty will be a major challenge. If this increased uncertainty is to be taken into account in life extension, the assessment should be based on significantly higher safety factors. However, at present no standards require such increased safety factors.

The history and present condition may be used to predict future degradation of the structure. Looking ahead, the evaluation should also include a risk analysis which covers future operations and which is updated with the relevant incident and accident history of the structure. Finally, any planned changes and modifications to the unit during the period of extended life should be included in the assessment.

Based on this information, the structure's integrity should be evaluated. The assessment should be performed with the aim of verifying whether the structure is sufficiently safe in its present condition, whether mitigations can be made to make it sufficiently safe or whether it should be decommissioned. However, the cost of mitigations may result in this being more an economical question than a structural safety question. If the decision is that the structure is sufficiently safe and it is possible to continue to use the structure, it is also important to establish an idea of the length of time the structure can be used safely. Further, it is important to establish which ageing mechanisms are likely to seriously threaten the safety of the structure and the symptoms of these ageing mechanisms. This evaluation can be used to identify possible inspection or monitoring activities in order to create good warning signals of future deterioration. The mitigating actions that are planned in order to provide the necessary level of safety may or may not work as intended. Hence, it is important to establish methods to evaluate the effect of these mitigations.

A life extension assessment also needs to result in an updated plan for structural integrity management (see Section 2.2), taking into account the ageing effects that the structure is assumed to be exposed to. A long-term inspection plan should normally be a part of this structural integrity management plan, as noted in Section 2.4.

The most general standard for offshore structures is ISO 19900 (ISO 2013). This standard gives general rules for both design and the assessment of existing structures. However, ISO 19900 does not give any specific requirements on how to perform an assessment. ISO 19900:2013 refers to ISO 19902 (ISO 2007) for the design and assessment of offshore steel structures. A detailed assessment procedure for existing structures is found in ISO 19902 and this procedure is presented in the next section. Other standards, such as NORSOK N-006 (Standard Norge 2015), API RP 2A-WSD (API 2014), and ISO/DIS 13822 (ISO 2000), also include detailed procedures for the assessment of existing structures. These are also presented in the following sections.

It should be noted that these standards use the word 'actions' (that includes loads but also includes all other influences that create stresses and strains in a structure) while in this book the word 'loads' is used.

4.2.2 Brief Overview of ISO 19902

ISO 19902 (ISO 2007) states that it is the owners' responsibility to maintain and demonstrate fitness for purpose of a platform for the given site and operating conditions. ISO 19902 clearly states that the design philosophy for existing structures allows for accepting limited damage to an individual component, provided that both the reserve against overall system failure and deformations remain acceptable. The standard is intended for application to existing jacket substructures but could also be used for topside structures.

The ISO 19902 procedure includes both a check of the ultimate limit state (ULS) and the fatigue limit state (see Section 2.1.2). Generally, if one of the platform assessment initiators in this standard exists, the structure shall undergo an assessment with several analytical and empirical methods available.

The first empirical method is to compare the structure with similar structures. The second empirical method is assessment by prior experience, where prior storm exposure that exceeds, by an appropriate margin, the loads the structure is required to withstand, is used as evidence of fitness for purpose. This requires knowledge of the structure having experienced these storms with no significant damage.

Analytically, the structure may be checked by the design formulae given in the ISO 19902 standard, calculating the explicit probabilities of structural failure or by evaluating the reserve strength ratio (RSR) of the structure.

Prevention and mitigation measures to reduce the occurrence rate and the consequences of structural failure should be considered for all cases that do not meet the design formulae in the standard, i.e. those that need to be evaluated by the empirical methods, by SRA or by RSR.

The triggers provided in ISO 19902 for an assessment to demonstrate fitness for purpose are similar to those given in Section 4.1.

The standard states that sufficient information should be collected to allow for an engineering assessment of the platform's overall structural integrity. Information of the platform's structural condition and facilities, with particular attention to data that cannot be explicitly verified (e.g. pile penetration), should be collected. General inspection of topside, underwater, splash zone and foundation should be performed and a decision on whether more detailed inspections and possible soil borings are necessary should be performed based on engineering judgement.

As noted above, a structure may, according to ISO 19902, be assessed by comparison with a similar nearby structure, where sufficient similarity can be demonstrated and the structure used for comparison is found to be fit for purpose. In ISO 19902, there is listed a series of requirements for the structure under assessment and for the similar structure for the comparison to be relevant and valid.

A non-linear analysis and component check is intended to demonstrate that a structure has adequate strength and stability to withstand a significant overload with respect to the applied loads. Local overstress and potential local damage are accepted but total collapse is not accepted. The ratio between the design loading (usually 100-year loading) and the collapse/ultimate capacity is then established and is usually referred to as the RSR in this book. The RSR shall be determined for all wave directions and the lowest value obtained is the structure's RSR. The criteria for acceptance of the RSR may differ depending on local requirements but ISO 19902 suggests an RSR acceptance criterion of 1.85 for manned structures (L1) and lower values of RSR are indicated as acceptable for unmanned structures (L2) (ISO 19902 section A.7.10.1). In ISO19902 section A.9.9.3.3, area specific RSR requirements are given for some areas, e.g. for Norway a RSR requirement of 1.92 is given for manned structures. However, the RSR type of analysis is not further referenced in the NORSOK N-series of standards in Norway.

Assessment for fitness for purpose may, according to ISO 19902, also be performed by a SRA. In ISO 19902, it is noted that the use of SRA requires extreme care and there is insufficient knowledge of the statistics to enable requirements or recommendations to be included in a standard.

If it can be demonstrated that a structure has already withstood events that are no less onerous than those for which it is to be assessed, the structure may be assumed to be acceptable. However, the prior event must be representative of all components and storm directions. It should be noted that the use of assessment by prior exposure provides limited results if the environmental loading has not exceeded the design environmental loading (see e.g. Ersdal 2005). It is likely that the environmental database should be revisited and this event should be included in the database and that a new analysis will show that this event no longer represents an extreme event.

Three different methods for assessing the structure with respect to fatigue are proposed in ISO 19902. According to ISO 19902, an 'extension of the design service life may be accepted without a full assessment if inspection of the structure shows that time-dependent degradation (i.e. fatigue and corrosion) have not become significant and there have been no changes to the criteria for design'. The three methods for assessment are:

- The results of a fatigue assessment show that the fatigue lives of all members and joints are at least equal to the total design service life (including an extension if needed) and the inspection history shows no fatigue cracks or unexplainable damage.
- A fatigue assessment has identified the joints with the lowest fatigue lives and appropriate periodic inspection of these joints has found no fatigue cracks or unexplained damage. This requires that the inspection that is carried out can be assumed to find relevant defects.
- Conservative fracture mechanics (FM) predictions of fatigue crack growth can demonstrate adequate future life and periodic inspection is able to monitor crack growth of the members or joints concerned.

4.2.3 Brief Overview of NORSOK N-006

The NORSOK N-006 standard (Standard Norge 2015) was developed to cover issues relevant for the assessment of existing structures primarily on the Norwegian Continental Shelf, and to be in line with the NORSOK N-series of standards. The fundamental requirements for the assessment of structural integrity are given in the NORSOK standard N-001 (Standard Norge 2012). NORSOK N-006 was developed to cover those aspects that are particularly relevant to the assessment of structures and issues of life extension.

The standard, together with the other NORSOK standards, is aimed to be a self-contained document but it has been developed to align with relevant ISO standards wherever possible. The standard covers all structural types of platforms and different types of structural materials but the emphasis is on steel jacket structures.

In contrast to the principles for life extension of existing structures given by API RP 2A-WSD (API 2014) and by ISO 19902 (ISO 2007), the recommendations in the NORSOK standard aim to ensure the same safety level for personnel as is required for new platforms.

The basic principles adopted for ULS and accidental limit state (ALS) checks (see Chapter 2) for existing structures are the same as for new structures. Load and material factors are therefore the same as those given in the other NORSOK standards. However, as the cost of implementing operational limits or structural reinforcements is large for an existing structure, additional requirements are given for how to perform advanced non-linear analyses for determining the structural strength. When non-linear analysis is used to determine the structural strength, the standard requires that low cyclic fatigue capacity is also checked. Hence, NORSOK N-006, includes a section on how to assess cyclic capacity for structures that are assumed to be used outside linear elastic behaviour.

Some important differences between the NORSOK N-006 standard and ISO 19902 and API RP 2A are that:

- NORSOK N-006 does not use the term RSR as a result of a non-linear analysis. Rather, it recommends using partial factors also in non-linear analysis. NORSOK N-006 is, as such, a fully partial factor method standard, even when non-linear analysis is used.
- Since the partial factor method is used in NORSOK N-006 also in non-linear analysis, the standard requires characteristic values of yield stress to be used in these

non-linear analysis. In contrast, ISO 19902 and API RP 2A indicate the use of mean values as is traditional in allowable stress design.

• NORSOK N-006 does not allow the use of SRA to document the safety of an existing structure.

Guidance is also given for fatigue assessment in NORSOK N-006. If the experienced service life for a structure is longer than the calculated fatigue life, it is indicated that it is possible to safely operate the platform further by using information about the performance and the inspection results. In addition, supplementary recommendations are given for fatigue analysis, acceptance criteria and improvement methods that are not given in other standards. Also, special recommendations for details that cannot be inspected are included.

4.2.4 Brief Overview of API RP 2A-WSD

API RP 2A-WSD (API 2014) is the most commonly used standard for the design of fixed offshore structures in the United States of America and many other areas of the world. The standard is stated to be applicable only for the assessment of platforms which were originally designed in accordance with the 20th or earlier editions of the same API standard. According to API RP 2A, structures designed after the 21st edition should be assessed in accordance with the criteria originally used for the design.

The elements of selection of platforms for assessment, categorisation of safety levels for the installation and condition assessment do not differ significantly from the ISO 19902 procedure.

There are two potential sequential analysis checks mentioned in API RP 2A-WSD: a design level analysis and an ultimate strength analysis. The analysis types suggested are more or less the same as mentioned in ISO 19902 but the acceptance criteria are different. Design level analysis procedures for assessment are similar to those used for new platform design, including the application of all safety factors. However, lateral environmental load may be reduced to 85% of the 100-year condition for high consequence platforms and to 50% for low consequence platforms. In the ultimate strength analysis, the RSR is defined as the ratio of a platform's ultimate lateral load carrying capacity to its 100-year environmental condition lateral loading. A RSR of 1.6 is required for high consequence platforms and a value of 0.8 for low consequence platforms.

In addition, assessment of similar platforms by comparison, assessment through the use of explicit probabilities of failure and assessment based on prior exposure are regarded as acceptable alternative assessment procedures subject, to some limitations similar to ISO 19902.

4.2.5 Brief Overview of ISO 13822

The three standards mentioned previously are all based on traditional structural analysis methods, such as semi-probabilistic partial factor methods and linear elastic design, whilst ISO 13822 (ISO 2000) is mainly a reliability based assessment standard. ISO 13822 lists land based structures in the scope of the standard and is not used particularly frequently for offshore structures. However, it is reviewed here for completeness.

The elements of selection of structures for assessment do not differ significantly from the ISO 19902 procedure. The objective of the assessment is specified in terms of the future performance required for the structure in an agreement with the owner, authorities and the assessing engineer. The required future performance should be specified in the utilisation plan and safety plan. Scenarios related to a change in structural conditions or actions should be specified in the safety plan in order to identify possible critical situations for the structure. The wording in this standard may differ from ISO 19902 but the elements are to a large extent similar.

ISO 13822 includes an option for a preliminary assessment and, if found necessary, a detailed assessment. The preliminary assessment includes verification of documents, a check of occurrence of large loads (actions), change in soil conditions, misuse of the structure and preliminary inspection for possible damage of the structure. ISO 13822 states that 'The preliminary inspection may clearly show the specific deficiencies of the structure, or that the structure is reliable for its intended use over the remaining working life, in which case a detailed assessment is not required. Where there is uncertainty in the actions, action effects or properties of the structure, a detailed assessment should be recommended'.

The detailed assessment, as described in ISO 13822, includes similar elements to ISO 19902, including detailed documentary search and review, detailed inspection and material testing, determination of actions, determination of properties of the structure and structural analysis. The degradation of the existing structure should be taken into consideration.

This being a standard based on structural reliability, it states that 'The verification of an existing structure should normally be carried out to ensure a target reliability level that represents the required level of structural performance'. In the informative part, it is further stated that 'The target reliability level used for verification of an existing structure can be determined based on calibration with the current code, the concept of the minimum total expected cost and/or the comparison with other social risks. The requirements should also reflect the type and importance of the structure, possible future consequences and socio-economical criteria'. Proposals for target failure probabilities for various limit states are also given. Several of these are not appropriate for offshore structures but the high and medium consequences of failure may be relevant. The respective indicated safety indices (β values of 4.3 and 3.8 represent a failure probability in the range 10^{-4}–10^{-5}).

An alternative approach to the structure intervention, which may be appropriate in some circumstances, is to control or modify the risk by imposing load restrictions, changing aspects of the use of the structure (e.g. demanning in severe storms, reduction of topside loading) and implementing monitoring and control regimes.

ISO 13822 also includes requirements for an assessment based on satisfactory past performance. These are quite restrictive, e.g. requiring that inspection has not revealed any evidence of significant damage, distress, or deterioration. Hence, in practice this is unlikely to be achieved for an offshore structure.

The client in collaboration with the relevant authority should make the final decision on interventions, based on engineering judgement and the recommendations in the report and considering all the information available.

4.2.6 Discussion of These Standards

The above review of relevant standards for the assessment of existing structures shows a number of common aspects but there are differences in the details. The choice of the standard to be used is often determined by regulators, and also in some cases by the operator's procedures and national practices. Many operators and regulators rely on ISO 19902 which was subject to a long period of development and review, with input from many stakeholders. The current version was issued in 2008 and, as the ISO standards are reviewed regularly, it is expected that ISO 19902 will be revised in the near future. In Norway and a few other countries, the NORSOK suite of standards is primarily used as the regulatory reference. These NORSOK standards are also subjected to a regular updating process. Overall, it is important that such standards reflect experience and recent knowledge in the relevant area and this is particularly important for the management of life extension as it is a relatively new process.

In most cases an assessment would first be performed using less advanced methods, such as linear analysis and code checks (similar to what is normally performed during design). However, if this fails to provide satisfactory results, the operator or engineers may choose more advanced analysis methods such as non-linear analysis or SRA. As noted previously, the use of non-linear analysis is rather complex and requires considerable expertise in its use. All failure modes, such as rupture, buckling, etc., have to be appropriately modelled and analysed and several non-standardised parameters such as the rupture criteria are often required. In addition, there is no agreement on how to implement safety factors in non-linear structural analysis.

The most common use of non-linear structural analysis for jacket structures is push-over analysis for the determination of RSR values of the structure. As shown above, ISO 19902 and API RP 2A have different requirements for the accepted value of RSR. In ISO 19902, the acceptable RSR is set to 1.85 for a high consequence structure, while in API the requirement is 1.6. This to some extent reflects the different regimes with respect to environmental conditions and level of manning on these structures. However, both standards are applicable for world wide use and care needs to be taken in applying API RP 2A to regimes with a harsh environmental climate. New knowledge indicates that the ISO 19902 requirement of an RSR of 1.85 may be too low (Ersdal 2005). It is shown in Ersdal (2005) that if the ratio of topside loading on a jacket (permanent and live load) to the environmental load on the jacket is not less than zero, a higher value of the RSR is required.

As an optional tool for assessing existing structures (in ISO 19902 and ISO 13288), SRA requires considerable specialist expertise which is at the time of writing only available in a limited number of companies. However, the use of SRA in inspection planning has proved to be very useful for structures during their design life. This type of analysis is commonly used by many companies. The use of such analysis, however, may be somewhat problematic for ageing structures (see Section 4.7).

API RP 2A seems to allow a lower safety level for structures undergoing assessment. This is exemplified by the allowance of the reduction of the lateral environmental load to 85% of the 100-year condition for high consequence platforms and to 50% for low consequence platforms. The authors consider that it is difficult to see any good reason for allowing such a reduced safety level for older platforms, especially if loss of life and environmental spills are possibilities.

4.3 Assessment of Ageing Materials

The main structural materials used offshore are steel and concrete and, for special purposes, aluminium (helideck and living quarters) and composite materials. Materials are specified at the design stage of an offshore structure, with properties based on those for new materials, and the design does not usually account for any degradation of properties due to ageing. Fabrication (e.g. welding) is also a key process during the early stages of the life of an offshore installation and degradation of the properties of a fabricated structure need also to be considered.

In an ageing structure, the possibility for a new material selection process and material replacement is limited when the installation is evaluated for life extension. In general, the operators have to accept the original materials. For some ageing processes, e.g. fatigue and corrosion, design lives are specified, such as that for fatigue life, which is set typically to 25 years.

There a number of degradation mechanisms that influence materials, which include corrosion metal loss, fatigue, hydrogen related cracking, wear and physical damage. All materials suffer some loss of performance as ageing occurs, which may have significance for structural safety. Ageing is a result of environmental effects on the materials, recognising that seawater is a particularly hazardous environment. The cyclic stresses that a material is subjected to can lead to loss of performance, particularly due to fatigue.

At the design and fabrication stage, data are required for the materials and processes being used. These include material certificates, welding procedures, results of non-destructive testing (NDT data), etc. At the life extension stage this data may no longer be available which introduces significant uncertainties in assessment.

Assessment of an existing facility for life extension involves a process where the operator needs to verify that the facility can be operated safely at acceptable risk levels. Failure modes and degradation mechanisms of the aged materials are important to identify, control and mitigate. It needs to be proved and documented that the materials selected in design, their fabrication and the quality of construction work are sufficiently robust to be fit for purpose also in the extended life.

Lange et al. (2004) divided the risk related to the selection of materials (for new designs) into four categories depending on uncertainty regarding materials and environment. In Hørnlund et al. (2011) this concept was further developed for assessment of material for ageing structures, see Figure 4.2.

Designing an offshore structure based on proven and known materials in well tested designs may be categorised as 'proven materials applied in known environments', according to the classification in Figure 4.2, which is the lowest risk level. Defining the technology to be within this categorisation implies that the operator has sufficient field data with respect to materials and design for the design life. However, lack of knowledge of aged and degraded materials can present an uncertainty in managing integrity. One opportunity to understand the change in performance of aged materials is from the testing of components from decommissioned structures, as has occurred in a few instances (Poseidon 2007). Test data and field experiences may be limited for new materials (e.g. composites) such that the 'materials knowledge' could be considered 'unproven', which is a higher risk level in Figure 4.2.

Where the environment has changed, for example as a result of subsidence, some parts of the structure will experience a new environment as parts of the facility that were originally above the splash zone are now in the splash zone, which results in heavier corrosion in areas not designed for corrosion. This change would bring a change to the risk levels, i.e. 'proven materials in a new environment'.

In the life extension stage, degradation mechanisms can make the material an unproven one eventually, based on the fact that data are not or are sparsely available for materials used in time frames representing the total service life of life extended facilities.

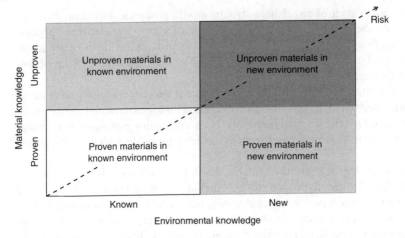

Figure 4.2 Risk and uncertainty in relation to robust materials selection. Source: Based on Hørnlund et al. (2011).

Hence, using a known structure with known materials beyond the original design life can in special circumstances become 'unproven materials applied in known environments'. The use of unproven materials in known environments is more challenging because general knowledge about the material has to be built up with respect to all actual failure modes and degradation processes of the material that may occur in the environment.

The class representing the most demanding challenge (and highest risk) is unproven materials applied in new environments. This is because there is limited general knowledge about the material and also of performance of materials in the environment. Even if there is information from laboratory testing and research work it may be incomplete as the various degradation processes have to be mapped. It has to be taken into account that the data may not be fully representative of actual offshore conditions. However, in terms of the structure there are very few situations now where unproven materials are operating in a new environment.

The characteristic parameters of a material are important in design to allow for any variation in properties, e.g. strength. Also, safety factors in the form of material factors are used to take uncertainty into account (see Section 2.2.2). Different materials have different safety factors, depending on the level of uncertainty in their properties. These are normally specified in codes and standards. However, the characteristic properties of the material and these safety factors may need to be reconsidered as a result of ageing. An understanding of the treatment of uncertainty relevant to life extension can be viewed via a 'traffic light' scheme as shown in Table 4.1. The three areas (green, amber and red) can be summarised as follows:

- *Green.* If the conclusions from the life extension evaluation are all within the green area the facility design is good, proven test data exist and the materials are fit for purpose for life extension.
- *Amber.* Some important data are missing and care will need to be taken in the determination of characteristic properties of the material and safety factors used in the assessment.

Table 4.1 Traffic light scheme for the assessment of ageing materials.

Green	Amber	Red
Material certificates present and verified, or substantial material testing performed	Materials certificates present for most elements.	Material certificates are lacking
Prediction of material degradation in whole life extension period is accepted with proven safety factors by corrosion and materials engineers	Limited evaluation of material degradation is performed and it is assumed that material degradation is acceptable for whole life extension period	No prediction of material degradation in life extension period is performed
Low level of material degradation, or intensive condition monitoring to ensure operation within design limitations	Medium level of material degradation, or limited condition monitoring	High level of material degradation, i.e. corrosion beyond design limitations. Minimum condition monitoring undertaken
Active use of inspection records in life extension assessment and operation	Inspection records are documented and reviewed, but not fully utilised in life extension assessment	Inspection records poorly documented or not used in life extension assessment
Utilisation of material information from testing and inspection of decommissioned installations. Lessons learned	Limited material information from testing and inspection of decommissioned installations	No assessment of external material data, particularly from decommissioned structures

- *Red.* Important data are missing and considerable care will need to be taken in the determination of characteristic properties of the material and safety factors used in the assessment for life extension.

The treatment of structural materials in the red or amber categories requires careful consideration of the safety factors during assessment, to ensure continued safety in the life extension stage. Indeed, for the red category, coupon testing in the field may be necessary to ensure continued performance as knowledge of the original selection of materials may be absent.

4.4 Strength Analysis

4.4.1 Introduction

The strength of a structure is its ability to withstand the applied load and load effects (such as stresses) without causing failure or causing a defined limit state to be exceeded. Assessment of the strength of existing structures is needed as part of a life extension process or due to other triggers that indicate that assessment is needed. This is partly due to possible damage or degradation defects that will reduce the load capacity of the structure but also if loads are found to be greater than those indicated in the original design analysis or if regulatory requirements have become more stringent.

The strength analysis of damaged and degraded structures and structural elements is one of the main challenges in the assessment of an ageing offshore structure. Various damage and degradation effects may be inflicted on offshore structures. These may include:

- corroded members
- cracked members and joints
- dents
- deflected members
- deformed shapes
- tears
- holes
- unusual deflections
- missing members
- wear
- hardening
- embrittlement

Each of these modes of degradation affects the strength of the structure in a different way and should be analysed by an appropriate technique. As an example, corrosion will first affect the thickness of a steel member and can be included by taking into account the metal loss and hence wall thinning in the calculation of section properties (e.g. area and moment of inertia). In addition, corrosion may introduce eccentricities in the member if the metal loss is unsymmetrical. This needs to be taken into account as an eccentricity may particularly influence the buckling capacity of a member. Further, corrosion may lead to more extensive fatigue cracking, which has to be taken into account in a fatigue analysis.

The effects of degradation mechanisms may be included in strength analysis by considering four major factors:

- Metal loss and wall thinning
- Cracking and partial removal of part of a section
- Changes to material properties
- Geometric changes

An overview of how the various degradation mechanisms may be included in the structural strength analysis is shown in Table 4.2. The assessment of the capacity of members needs to take these ageing effects into account, as further discussed in this section.

4.4.2 Strength and Capacity of Damaged Steel Structural Members

The most likely forms of degradation and damage to an offshore steel structure are corrosion, fatigue cracks, wear, dents and buckling. Some studies also indicate that the elastic limit is reduced in severely corroded samples, indicating that the structure becomes more brittle with increasing corrosion (Saad-Eldeen et al. 2012). These need to be taken into account in the strength calculation, the ULS and ALS.

The most common modes of failure of steel members are due to excessive stresses due to bending, axial and shear loads (or a combination of these), bearing failure and local and global buckling. The first four of these failure modes are primarily affected by the

Table 4.2 Ageing effects and the effect on the structures.

	Metal loss and wall thinning	Cracking and removal of part of section	Changes to material properties	Geometric changes
Corrosion	X	X	X (see Saad-Eldeen et al. 2012)	X
Cracking		X		X
Denting				X
Deformed shapes				X
Tears		X		
Holes		X		X
Wear	X			X
Hardening			X	
Embrittlement			X	

fact that the section area and other section properties are reduced by the material loss as a result of corrosion, fatigue cracks and wear. In addition, a reduction in yield stress will also affect these failure modes. Local and global buckling will also be affected by any eccentricity introduced by damage and degradation.

4.4.2.1 Effect of Metal Loss and Wall Thinning

The primary effect of metal loss and wall thinning is the reduction of section properties such as area, section modulus and moment of inertia. The axial, shear and bearing capacity will be dependent on the section area (or the parts of the section area that are able to carry these loads), and any reduction in the section area will reduce the capacity of the steel beam with respect to these loads. Moment and buckling capacity will similarly be affected by changes to the section modulus or moment of inertia.

If the metal loss and wall thinning is unsymmetrical, locally or along the beam, the metal loss and wall thinning may introduce eccentricities, as discussed later as a geometric change.

An important aspect of the strength of a steel member is the ability to deform plastically or elastically prior to local bucking occurring. Steel beams will often be assessed in four classes depending on their failure mode. Class 1 will typically be steel beams that are able to fully develop plastic hinges with the necessary rotation capacity without reduction of the resistance prior to failure. Class 2 beams can develop plastic moment resistance but have limited rotation capacity as they may experience local buckling. Class 3 beams have cross sections where the capacity can be calculated by elastic methods and will not buckle locally prior to yielding in the extreme compression fibre. However, local buckling must be expected if loading exceeds this level. Finally, Class 4 beams have slender cross sections that will experience local buckling prior to yielding in the extreme fibre. The behaviour of these four classes of sections is shown in Figure 4.3.

A structural steel member that has experienced wall thinning by corrosion or other degradation mechanisms may have to be reclassified based on the new thicknesses of the section. Even if the section properties do not change significantly, a reclassification of the section may lead to a relatively significant drop in the section's ability to withstand moment, axial, shear and bearing loads.

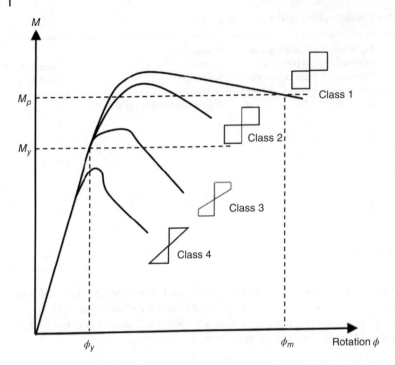

Figure 4.3 Section classes for beams. M_y, elastic moment capacity; M_p, plastic moment capacity.

4.4.2.2 Effect of Cracking and Removal of Part of Section

The effect of cracking and removal of part of the section (e.g. due to boreholes, extensive damage, etc.) will normally be both a reduction of section properties and geometric changes by the introduction of eccentricities.

4.4.2.3 Effect of Changes to Material Properties

Material properties may change due to hydrogen embrittlement (see Section 3.3.3.2), hardening of materials (see Section 3.3.3.1) and possibly also by corrosion. If non-linear plastic analysis is used (or was used at the design stage), the ability of various material properties, e.g. the rupture stress, to take plastic strain and hardening may need to be updated as a result of degradation.

4.4.2.4 Effect of Geometric Changes

Most degradation mechanisms will introduce geometric changes, typically by introducing eccentricities to the structural member. Severe dents, corrosion, cracking, etc. will disturb the geometry and centroid of the section and hence introduce eccentricities that may influence in particular the buckling capacity of the section, beam, or plate.

An illustration of degradation introducing eccentricity is shown in Figure 4.4.

4.4.2.5 Methods for Calculating the Capacity of Degraded Steel Members

A number of research reports have been issued on the strength of damaged bracing members, typically dented tubular braces (e.g. Smith et al. 1979; Smith et al. 1981; Ellinas and Walker 1983; Taby and Moan 1985; Smith 1986; Yao et al. 1986; Taby and

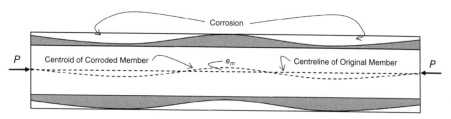

Figure 4.4 Worst-case external asymmetric corrosion. Source: Based on Lutes et al. (2001).

Moan 1987; Landet and Lotsberg 1992). These research reports have culminated in the requirements in standards such as ISO 19902 (ISO 2007), NORSOK N-004 (Standard Norge 2013) and API RP 2A (API 2014) which all give guidance on how to calculate the capacity of degraded tubular structural steel members normally used in jacket structures. In these standards, the effect of the geometric change due to the dent and member bowing is taken into account and includes the out of straightness that will typically occur in denting and bowing events.

Corrosion and cracks have in these standards been modelled as an equivalent dent. As an example, the equivalent dent to represent a crack or corrosion was in the 1998 version of NORSOK N-004 suggested to be:

$$\delta = \frac{1}{2}\left(1 - \cos\left(\pi \cdot \frac{A_{crack}}{A_0}\right)\right) \cdot D; \delta = \frac{1}{2}\left(1 - \cos\left(\pi \cdot \frac{A_{corr}}{A_0}\right)\right) \cdot D$$

where δ is the equivalent dent due to a crack, A_{crack} is the area of the cracked part of the cross section, A_{corr} is the area of the corroded part of the cross section, A_0 is the cross-sectional area of the undamaged or uncorroded member and D is the outside diameter of the undamaged member. An example of the reduction of the capacity of a dented tubular member is shown in Figure 4.5 based on the use of the 1998 version of NORSOK N-004. It shows a 20% reduction in capacity for a dent of only 30 mm (or 10% of the section corroded) and a 50% reduction for a dent of only 85 mm (or ~17% of the section corroded). These dents and corroded sections are difficult to find during inspection, particularly when marine growth is involved. Hence, for ageing structures where many such dents and corroded sections may exist, the effect on the capacity of the structure may be significant and not necessarily recognised.

DNVGL-CG-0172 (DNVGL 2015b) gives guidance for calculating the capacity of degraded steel members in ship-shaped structures, semi-submersible structures (column-stabilised units) and jack-ups. Ship structures and other floating offshore steel structures are normally reassured by the capacity of their hull girders, individual panel buckling checks and minimum requirements for local scantlings. Classification rules and some standards also give some indications of how to perform assessment calculations. However, research is ongoing on the topic and, for example, Saad-Eldeen et al. (2013, 2015) indicate a significant reduction in capacity of a hull girder with increasing levels of corrosion. Saad-Eldeen et al. (2013, 2015) indicate a reduction in capacity of a severely corroded member by a factor of the order of three.

4.4.3 Strength and Capacity of Damaged Concrete Structural Members

As noted in Section 3.4.5, the most likely forms of damage to an offshore concrete structure are loss of cover to the reinforcement, corrosion of the reinforcement, accidental

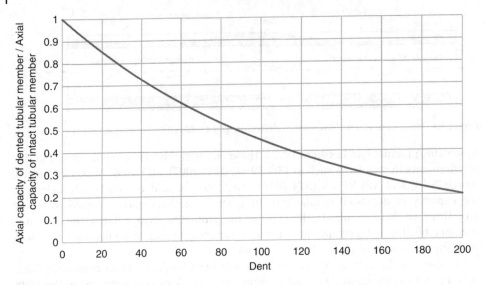

Figure 4.5 Reduction in axial capacity of dented tubular member (D=1.2 m; t=10 mm).

damage to the concrete from either dropped objects or ship collision and corrosion of prestressing tendons. Concerns about such damage are associated with the effects on strength and the potential water ingress leading to corrosion or loss of operational capability. Accidental damage to the concrete legs can cause leakage, leading to loss of pressure control in the cells and consequent overstress of the cell/legs connections or direct loss of strength in the area of the damage.

Local loss of concrete cover to the reinforcement are likely to result in minimal loss of strength. However, failure to repair this loss can lead to more severe local corrosion of the reinforcement but this unlikely to affect the overall strength significantly at a local level.

Accidental damage to the domes of storage cells from dropped objects can lead to both loss of strength and seawater containment. This has happened in a few cases and significant damage needed repair. In one case (Ocean Structures 2009), damage to a 500 mm thick cell roof slab resulted from a dropped object, causing a deep hole 300 mm in depth which led to water flowing though the slab. Repair was undertaken by pre-packing aggregates within the hole, covering the hole with a steel plate and injecting grout to restore the original concrete profile.

Damage to the concrete cover on the legs from ship impact is also likely to need repair, particularly for limiting future reinforcement corrosion. An example of ship impact quoted in Ocean Structures (2009) resulted in water seeping through a cracked wall. It was repaired using resin and caulking to seal the damaged area externally and a section of wall was then removed internally and recast.

In many concrete platforms the stresses in the base cells and the adjacent parts of the leg are influenced by the internal pressure in the cells and loss of the pressure regime within the cells can lead directly to possible overstress of the structure. This pressure is controlled by the water level in a ballast header tank within one of the legs. The normal underpressure of the cells is approximately three to four bar in relation to the adjacent sea water. Any failure of the ballast water system or water ingress to the cells as a result of damage can lead to a loss of control of the ballast water system with possible overstress of the structure.

Corrosion of prestressing tendons in ducts which have not been fully grouted could lead to loss of strength, depending on the extent of the problem, as noted in Section 3.4.5. A study commissioned and reported on by the Health and Safety Executive (HSE 1997) investigated the potential for loss in strength of a typical offshore concrete structure due to local loss of prestressing resulting from corrosion. The study also reviewed different methods for inspecting prestressing tendons. The analyses undertaken for a typical structure in the report showed that there would have to be a 10% loss of the prestress to cause the reliability to fall below the target value and 40% before a leg could fail under loading from a design wave. In addition, it would be necessary for all the tendons to fail in the same section which is unlikely for grouted ducts. However, as noted in a report on land based structures (Tilly 2002), there has been evidence where failures have occurred in the same section, mainly as a result of failure at anchorages. In offshore structures, the anchorages of vertical tendons may be more susceptible to inadequate grouting than for the horizontal post-tensioned beams used in bridge decks. As these vertical tendons play an important part in limiting bending of the legs due to wave loading, inspection of the anchorages is important and repair is needed if damaged.

The report (Tilly 2002) also identified the most vulnerable area in an offshore concrete structure as the steel to concrete transition between the deck and platform legs. It is recommended in the report that inspection should be made of this location as failure could lead to structural damage.

The report also notes that, as structures reach their design life and are considered for life extension, it would be necessary to assess their structural condition. This includes inspection of the prestressing systems so that they can be confirmed as being in a satisfactory condition. However, inspection of these systems is not easy as has been identified for bridge type structures (Tilly 2002).

4.4.4   Non-Linear Analysis of Jacket of Structures (Push-Over Analysis)

As mentioned earlier, non-linear analysis of structures may be a reasonable approach to determine the strength of ageing structures if linear analysis fails to give acceptable results. In practice, non-linear analysis has been used mainly for jacket structures.

Theories for the non-linear collapse (pushover) analysis of jackets have been established by Søreide and Amdahl (1986). The approach was further developed by Hellan (1995) and Skallerud and Amdahl (2002). A guideline for performing non-linear collapse analysis is given in the Ultiguide project (DNV 1999).

Non-linear pushover analysis may be performed in a limit state and partial factor format, where the ultimate strength (or ultimate collapse capacity) R_{ult} is calculated based on characteristic values of strength for the characteristic values of loads. The limit state check is then performed as:

$$\left(\sum Q_i \cdot \gamma_i\right) \cdot \gamma_m \leq R_{ult}$$

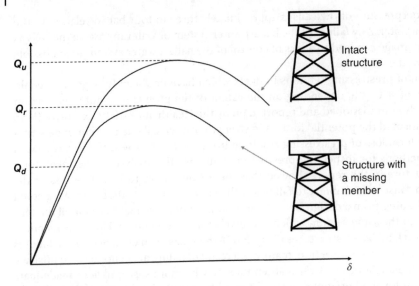

Figure 4.6 Illustrative $Q-\delta$ curve for a jacket structure with indication of the design load level (Q_d) and collapse load level in intact (Q_u) and damaged (Q_r) situations, where Q along the vertical axis is load and δ along the horizontal axis is deformation.

where Q_i is the load i, γ_i is the partial factor for load i and γ_m is the material factor. An additional safety factor for non-linear analysis is often recommended due to the uncertainties introduced in advanced analysis, e.g. DNVGL-RP-C208 (DNVGL 2016b). In practice, since this is a non-linear analysis and the superposition principle cannot be used, the loads have to be added into the analysis one by one. For a jacket-type structure, the weight of the structure and topside will typically be added first, then the live load if relevant and finally the wave load.

Alternatively, the RSR approach can be used. In this approach, the safety requirement is specified as the ratio between the design load and the collapse capacity of the structure with the given loading distribution. The ratio between the design loads (typically the characteristic load) and the collapse capacity of the structure is established in the collapse analysis and is normally called the RSR. An illustrative $Q-\delta$ curve (load versus deflection curve) is shown in Figure 4.6 for both an intact structure and a damaged structure. RSR is defined as:

$$RSR = \frac{Q_u}{Q_d}$$

where Q_u is the load that ultimately results in collapse of the structure and Q_d is the design load for the structure.

In addition to the RSR, a similar parameter giving the ratio between the collapse capacity of the damaged structure and the design load is used and is often referred to as the damaged strength ratio (DSR). This factor gives an indication of the redundancy of the structure, i.e. the ability of the structure to survive without a particular member. With reference to the load levels in Figure 4.6, the DSR can be defined as:

$$DSR = \frac{Q_r}{Q_d}$$

where Q_r is the load that ultimately results in collapse of the structure in the damaged condition.

The reference level for the design load used in the RSR and DSR calculations would typically be the characteristic wave, current and wind loading (the load with an annual probability of 10^{-2}). In the RSR and DSR calculations, the design load does not include any load factor.

Typical values of the required RSR range from 1.6 to 2.4 for critical structures. The type of framing selected at the design stage can have a significant influence on the reserve strength of the system. This parameter is a factor in designing inspection plans for ageing structures and a high RSR value may compensate for limited inspection on a global basis.

An additional factor illustrating the loss in strength when a component is damaged or missing is given by the residual strength factor (RSF). The RSF is defined as:

$$RSF = \frac{Q_r}{Q_u} = \frac{DSR}{RSR}$$

Hence, the RSF is a measure of the effect on the RSR when a member is damaged or lost and, as such, is a good indication of the redundancy in the structure. If the RSF values are close to 1.0 for all members, the structure can be considered to have relatively good redundancy.

4.5 Fatigue Analysis and the S–N Approach

4.5.1 Introduction

An overview of fatigue and its significance to ageing structures was presented in Chapter 3. It is apparent that offshore structures are vulnerable to fatigue due to the effects of cyclic loading, especially from waves, which is amplified at welded joints whose geometry introduces significant SCFs. Fatigue crack initiation can occur under high cyclic stresses acting on welds containing microscopic defects which are inherent due to the welding process. The likelihood of crack initiation and propagation increases in ageing structures as fatigue is a time-dependent process. Thus, the effective management of ageing and life extension needs to take full consideration of the effects of fatigue using analytical methods to determine fatigue life predictions and an appropriate inspection strategy.

The cyclic stresses acting at welded joints, arising from the stress concentration effects of the structural and weld geometry on the applied loading, need to be taken into account in the determination of the fatigue life. The stress distribution at a welded joint is depicted in Figure 4.7. The nominal stress (S_{nom}) is the stress in the member without any influence of the geometry of the welded connection and is used in the nominal stress method for fatigue analysis, as described below. A hot-spot stress (S_{hot}) is a local stress at the hot spot where cracks may be initiated. It accounts for the stress concentration due to the influence of structural geometry of the connection and is also referred to as structural stress. It is found by linearly extrapolating the stress at $3/2t$ and $1/2t$ away from the weld toe to the weld toe, as shown in Figure 4.7. This stress is used in the hot-spot stress method for fatigue analysis, as further described below. The notch stress (S_{notch}) is the peak stress at the weld toe or notch taking into account stress

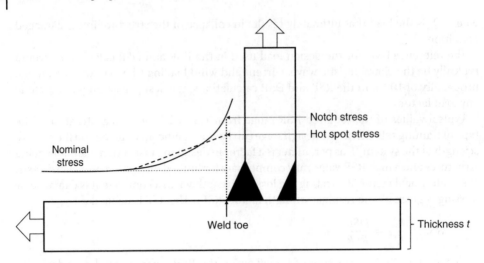

Figure 4.7 Stresses in a welded connection in a structure.

concentrations due to the effects of structural geometry as well as the presence of the weld. This stress is used in the notch stress method of fatigue analysis and in fracture mechanics assessments (see Section 4.6).

Fatigue analysis methods are based on the assumption that cracks will initiate under cyclic loading, the number of cycles depending on the applied loading. The initiation stress/cycles are also sensitive to the size of weld microscopic flaws not detected by NDT during fabrication, emphasising the significance of weld quality to fatigue life. The crack propagation rate and consequently the fatigue life are determined by the cyclic stresses and the crack size. Both empirical and theoretical techniques based on stress analysis of structures containing defects have been developed for the prediction of fatigue life.

4.5.2 Methods for Fatigue Analysis

There are two main methods of fatigue analysis for the assessment of the life of a structure under cyclic loading before failure by crack propagation occurs, namely the *S–N* and the fracture mechanics approaches, both of which are relevant to the design operation, and life extension of offshore structures.

For fatigue design purposes, the *S–N* approach, which is based on *S–N* curves determined from experimental data, is widely used and is more suitable.

The fracture mechanics method is used to predict the fatigue life based on the propagation (growth) of initial defects or a defect found during inspection of the structure. Fracture mechanics can further be used to determine limiting flaw sizes and to plan inspection and repair strategies and is hence useful in life extension assessments. Fracture mechanics analysis enables more detailed assessment of the fatigue life when the *S–N* approach predicts insufficient fatigue life, as fracture mechanics rely on the assumption of an initial crack size based on measurement from inspection or on NDT limits (typically the crack size with 90% probability of detection for the given inspection method used).

The following conditions and parameters are important in the quantification of the fatigue life:

- Cyclic loading (relevant for both *S–N* fatigue and fracture mechanics assessment).
- Geometry of the welded detail (relevant for both *S–N* fatigue and fracture mechanics assessment).
- Environmental condition during service and the presence of corrosion protection systems (relevant for *S–N* fatigue in selection of the *S–N* curve and for fracture mechanics assessment).
- Material characteristics (important for fracture mechanics assessment, less so for *S–N* assessment).
- Misalignments and eccentricities (important for fracture mechanics assessment and relevant for *S–N* fatigue if the misalignment exceeds the amount which is already implicit in *S–N* curves for the structural detail).
- Residual stresses (important for fracture mechanics assessment, less so for *S–N* assessment as these are implicitly included in the *S–N* curves via the R-ratio).
- Constant stresses from constant loading such as weight and permanent loading (relevant to *S–N* fatigue and fracture mechanics assessment but rarely used in *S–N* fatigue).
- Production quality and surface finishing (relevant to both *S–N* fatigue and fracture mechanics assessment).

4.5.3   *S–N* Fatigue Analysis

The *S–N* approach is the traditional method of fatigue life assessment and is based on the use of *S–N* curves in conjunction with a long-term fatigue stress range distribution or spectrum providing the number of fatigue cycles (N) for each stress range (S). A considerable amount of effort has been expended in recent decades to generate *S–N* curves in general and specifically for offshore structural components.

The basic principle of the *S–N* fatigue approach is shown in Figure 4.8.

The steps in this process are described below.

4.5.3.1   Fatigue Loads and Stresses to be Considered

All types of fluctuating and static loads acting on a component and the resulting stresses at potential sites for fatigue, which are determined in accordance with the selected fatigue assessment procedure, have to be considered. The stresses in a structure originate from live loads, dead weights, snow, wind, waves, pressure, accelerations, dynamic response and transient temperature changes. Inadequate knowledge of fatigue loads and stresses is a major source of uncertainty and inaccuracy in fatigue life predictions.

The most important fatigue loading on an offshore structure is the global cyclic loading due to waves. On average, every 8–10 seconds a wave will force a structure through a full cycle of loading in the wave direction and against the wave direction. This cyclic loading from waves will also induce motions and accelerations in the rest of the structure which will also give cyclic loading; especially in floating structures and highly dynamic structures these cyclic loading are important. Waves may also give slamming loads on the structure (e.g. on braces in the splash zone and plates in floating structures). In special cases, the members and plates may vibrate due to this slamming load and may accumulate damage from this loading.

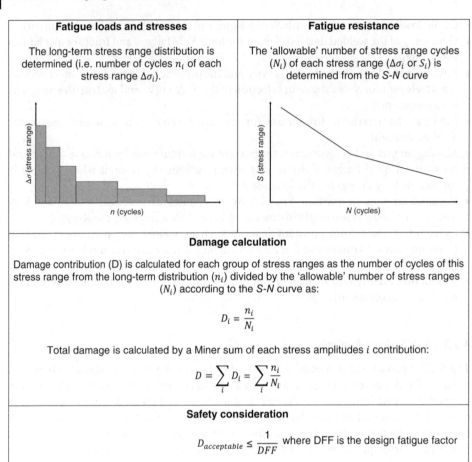

Figure 4.8 *S–N* approach calculation flow diagram (simplified).

Wind is also an important fluctuating load that may induce cyclic loading on structures above water. The two main loads from wind contributing to fatigue are fluctuating gust wind and vortex shedding. Fluctuating gust wind is most relevant for dynamic global structures while vortex shedding is most relevant for dynamically sensitive single members. Current may also induce vortex shedding and hence fatigue in some cases.

In addition to the in-place cyclic loading, the transport and installation phases contribute to this loading. During transportation, structural members are exposed to cyclic accelerations and may be exposed to vortex shedding due to wind. During installation, pile driving induces significant cyclic loading on the piles.

Residual stresses due to welding can also have a sizeable influence on the fatigue life. Tensile residual stresses decrease the fatigue resistance because they raise the mean stress, whilst compressive stresses improve fatigue resistance by reducing the mean stress and hence reducing the tensile stress range.

There are four methodologies for calculating the long-term stress range distribution from global wave loading:

- Deterministic discrete wave analysis.
- Simplified fatigue analysis (also known as closed form fatigue analysis).
- Spectral fatigue analysis.
- Time domain fatigue analysis.

These methods are described in DNVGL-RP-C210.

The effects of misalignment (eccentricity) on the secondary bending stress caused by axial or angular misalignment should be considered.

4.5.3.2 Fatigue Capacity Based on S–N Curves

There are three fundamental methodologies for the evaluation of fatigue stress with particular S–N curves for each method:

- The nominal stress method.
- The structural hot-spot stress (SHSS) method.
- The notch stress method.

The first two methods are the most commonly used and these are discussed below in more detail.

Nominal Stress Method for the S–N Approach   The nominal stress approach is based on extensive tests of standardised welded joints and connections. The joints are classified by type, loading and shape. It is assumed and confirmed by experiments that welds of a similar shape have the same general fatigue behaviour so that a single design S–N curve can be employed for any connection in this weld class. The analysis can be based on the nominal stress and any stress concentration effects of the detail and weld are implicitly included in the S–N curves. However, stresses arising from the effect of the macro-geometric shape of the component in the vicinity of the joint, such as large cut-outs that are not included in the weld class and the effect of concentrated loads should be included.

The nominal stress method entails the following steps:

1. Selection of weld class and corresponding S–N curve, as defined in standards (e.g. ISO 19902, DNVGL-RP-C203).
2. Identification of environment (air, seawater with cathodic protection seawater with free corrosion) for selection of the S–N curve.
3. Calculation of nominal stress ranges.
4. Correction of stress range for the thickness effect and misalignment (if relevant).
5. Determination of the number of cycles to fatigue failure from the S–N curve.
6. Damage calculation and safety assessment, as shown in Figure 4.8.

As the nominal stress is the mean cross-sectional stress without any influence of the geometry of the welded connection, it can be determined using elementary theories of structural mechanics based on linear-elastic behaviour.

It should be noted that geometric features in the vicinity of the weld, such as cut-outs and transitions, concentrated forces and reaction forces, can influence the stress field beyond what is included in the weld class. In order to obtain the nominal stress in such cases, finite element method (FEM) modelling may be used. Using the FEM, meshing can be simple and coarse. However, care must be taken to ensure that all stress raising effects of the structural detail of the welded joint are excluded when calculating the modified (local) nominal stress.

Hot-Spot Stress Method Fatigue design of more complicated details, where a standardised weld class is not available or where the dimensional variation of a particular detail has to be taken into account, is normally based on the hot-spot stress method. The SHSS incorporates the geometrical stress concentration which is more representative of the stresses experienced. The hot-spot stress is the local stress at the weld toe, taking into account the overall geometry of the joint excluding the shape of the weld in Figure 4.7. In recent code versions, the stress at the weld toe is extrapolated from two or three points near the weld toe, e.g. as recommended in DNVGL's RP-C203 (DNVGL 2016a).

The hot-spot stress to be used in the $S–N$ approach calculation is normally given as the nominal stress multiplied by a SCF:

$$\Delta\sigma_{HotSpot} = SCF \cdot \Delta\sigma_{nominal}$$

However, the hot-spot stress may also be calculated directly, for example by FEM analysis.

The hot-spot stress method is used for tubular structures and parametric SCF equations are given for simple tubular joint configurations (see, for example, Efthymiou 1988). These more modern methods for calculating SCFs for tubular joints use a generalised influence function approach that differentiates between the types of loading in the joint. It should be noted that SCF equations for tubular joints have evolved over the years. Many older structures have been designed with the limited information available at the time which in the early years relied mainly on the Alpha-Kellogg equations, see Kinra and Marshall (1980), and did not reflect the many combinations of the geometric and loading parameters. These equations are considered to generally underpredict the joint SCF. Considerable research and development work followed, resulting in extensive parametric equations for offshore tubular joints. It follows that the reassessment of such structures using SCF equations in current standards are significantly more advanced and are likely to generate significantly different fatigue life predictions, which may be lower than the original predictions.

Design S–N Curves Design $S–N$ curves have been developed for use in the nominal stress method, the hot-spot stress method and the notch stress method.

$S–N$ curves for the nominal stress approach are divided into several classes depending on the geometry and the direction of the fluctuating stresses in the welded joint. Each class has a designated $S–N$ curve. Each construction detail, at which fatigue cracks may potentially develop, should be placed in its relevant joint class in accordance with criteria given in the codes (e.g. DNVGL-RP-C203). Fatigue cracks may develop at several locations, e.g. at the weld toe in each of the parts joined, at the weld root, and in the weld itself. Each location needs to be classified separately in accordance with the guidelines in codes, standards and guidance. The types of joint, including plate-to-plate, tube-to-plate and tube-to-tube connections, have alphabetical classification types, where each type relates to a particular $S–N$ relationship as determined by experimental fatigue tests. For example, Norwegian and British codes reference the D curve for simple plate connections with the load transverse to the direction of the weld.

In the hot-spot stress approach, S–N curves that do not include the effect of the geometric stress concentration should be used. In DNVGL-RP-C203, the D curve is recommended. The fatigue analysis of tubular joints in jackets is normally performed using the hot-spot stress approach in conjunction with the use of parametric equations for SCFs. A specific T curve has been developed for tubular joints and included in most fatigue standards for offshore structures (e.g. DNVGL-RP-C203).

The design S–N curves are based on characteristic values, i.e. the mean-minus-two-standard-deviation curves, for relevant experimental data. This is analogous to the approach used for the characteristic strength of the material. Thus, the S–N curves are associated with a 97.7% probability of survival.

The basic design S–N curve is given as:

$$\log N = \log A - m \log S$$

where S is the stress range, N is the predicted number of cycles to failure for stress range S, m is the negative inverse slope of the S–N curve (typically $= 3$) and $\log A$ is the intercept of the $\log N$ axis on the S–N curve.

S–N curves normally have a bi-linear relationship between $\log S$ and $\log N$ and the change in slope from a gradient of 1/3 to a gradient of 1/5) occurs typically at N in the range of $10^6 - 10^7$ cycles. The lower right side of the S–N curves reflects the considerably longer life determined from tests on joints at low stress ranges.

Effect of Environment on the Choice of S–N Curves   Design S–N curves for structures in sea-water with corrosion protection are given in the various offshore structural codes, e.g. ISO 19902:2007 and DNVGL-RP-C203 (DNVGL 2016a). In general, the fatigue life in seawater under free corrosion is approximately one third of the life in air at high stress ranges (when N is less than 10^7 cycles).

The data for fatigue with cathodic protection (CP) demonstrate clearly that, whilst the fatigue life is lower than that in air, the effect of CP is to improve the fatigue life substantially compared with that under free corrosion. This requires that the level of CP is maintained at the appropriate range (as described in Section 3.4) and it is apparent that the maintenance of CP systems is an important aspect of the fatigue life management of offshore structures, particularly where life extension is concerned. Indeed, the failure to maintain the CP system can contribute to the ageing of the structure.

4.5.3.3   Damage Calculation

The S–N curves are primarily based on the testing of specimens exposed to cyclic stresses at one stress level. In real life, structural elements, and specifically offshore structures, are subjected to varying stresses. The S–N curves for offshore structures have been derived from data obtained under variable amplitude loads (which are represented by an effective stress range) to simulate the offshore environment. The Miner (or Palmgren–Miner) summation is the most commonly used method to determine cumulative damage. The rule is based on the assumption that the total damage accumulated is obtained by the linear summation of the damage of each individual stress range, given by:

$$D = \sum_i D_i = \sum_i \frac{n_i}{N_i}$$

Table 4.3 ISO 19902 fatigue safety factors.

	Critical	Non-critical
Accessible	5	2
Not accessible	10	5

Table 4.4 NORSOK N-001 fatigue safety factors.

Classification of structural components based on damage consequence	Not accessible for inspection and repair or in the splash zone	Accessible for inspection, maintenance and repair, and where inspections or maintenance are planned	
		Below splash zone	Above splash zone or internal
Substantial consequences	10	3	2
Without substantial consequences	3	2	1

where D is total cumulated damage, n_i is the number of cycles of constant amplitude stress ranges $\Delta\sigma_i$ and N_i is the total number of cycles to failure under constant amplitude stress ranges $\Delta\sigma_i$.

4.5.3.4 Safety consideration by Design Fatigue Factors

A safety margin is introduced in the $S–N$ approach through the application of design fatigue factors (DFFs), the value of these depending on the criticality and the accessibility of the structural component being assessed. Design fatigue factors range from 1 to 10 in ISO 19902 (ISO 2007), NORSOK N-001 (Standard Norge 2012) and the American Bureau of Shipping (ABS) document 'Guide for Fatigue Assessment of Offshore Structures' (ABS 2014). The safety factors in ISO 19902 are shown in Table 4.3 and those in NORSOK N-001 are given in Table 4.4.

The ISO 19902 DFFs represented a significant increase compared with the numbers that had been used previously (typically a factor of 2). Further work was carried out in standardisation committees to verify these factors and NORSOK N-001 was calibrated with SRA (Moan 1988).

4.5.4 Assessment of Fatigue for Life Extension

4.5.4.1 Introduction

The codes, standards and recommended practices ISO 19902, NORSOK N-006 and the ABS document 'Guide to Fatigue Assessment of Offshore Structures' (ABS 2014) address fatigue life in relation to life extension, both for fixed platforms and for mobile installations. As already mentioned, the basic approach is through the application of design factors to fatigue life. It should be noted that older designs of the 1970s and 1980s had lower design factors (often based on a factor of 2) than what is the practice today.

The DNVGL rules for the classification of offshore drilling and support units require that when the life exceeds the design fatigue life, mobile installations should be subjected to special evaluation prior to the renewal survey when the nominal age exceeds the documented fatigue life. The fatigue utilisation index (FUI) is determined and is defined as the ratio between the effective operational time and the fatigue design life, i.e. the ratio of the actual consumed fatigue capacity to the design life.

If no cracks are found prior to the FUI reaching a value of one, no special provisions are required until such cracks are found. If fatigue cracks are found prior to the FUI reaching one, the owner is required to assess the structural details in the relevant areas prior to the renewal survey for the five-year period in which the FUI is assessed to reach one. A key purpose of assessment is to improve the fatigue properties of the structure (e.g. by replacement or grinding out of defects).

In cases where units have suffered fatigue cracking prior to the FUI reaching one and where satisfactory compensating measures in the form of structural improvements have not been implemented, the codes recommend that these units should be subject to additional non-destructive examination (NDE) at intermediate surveys corresponding to the much larger extent required for the renewal survey. An additional requirement is for an approved leak detection system to be fitted when the FUI exceeds one and areas identified for leak detection. When the FUI exceeds one, potentially vulnerable areas should be examined for leaks at least twice-monthly and consideration given to additional measures, including a fracture mechanics crack growth assessment of the time to failure of the largest through-wall crack that could escape detection from leakage.

ISO 19902 states that for structures that have their life extended or are re-used or converted to a new application, prior fatigue damage may have to be estimated via inspection findings. An absence of crack discoveries should *not* be assumed to mean no prior damage has occurred. It is assumed the the prior damage in terms of the Miner's sum is 0.3 for a welded tubular joint and 0.5 for a welded plate detail. These figures were derived in a study by Tweed and Freeman (1987) which assessed initiation lives for a large number of tubular and plate joints and represent average estimates of the proportion of the initiation life to the overall fatigue life. The standard also considers that, assuming a defect-free inspection, lower values of assumed damage may be justified by the designer. This justification may be based on analysis if the prior history of the structure can be established with confidence. However, a value of zero is usually only used for those details that will be modified so as to eliminate prior damage. The ABS document 'Guide to Fatigue Assessment of Offshore Structures' (ABS 2014) has a similar rule to cover situations where an existing structure is being re-used or converted.

4.5.4.2 High Cycle/Low Stress Fatigue

Ageing structures are exposed to large numbers of low stress cycles and the prediction of the fatigue life is determined, amongst several factors, by the characterisation of the $S–N$ curve for low stress ranges. The $S–N$ curves have a change in slope, as shown in Figure 4.9, and the point which the change in slope occurs has been the subject of considerable debate in the course of the development of the codes. The paucity of data due to the considerable time and cost associated with undertaking such tests is a significant impediment to the generation of sufficient results to reduce the uncertainty of the $S–N$ performance. A change in slope at N_0 cycles was adopted in the HSE fatigue guidance, as described in the background document (HSE 1999), and subsequently in ISO 19902 and

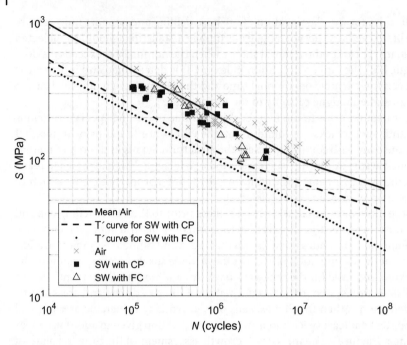

Figure 4.9 *S–N* data for tubular joints in seawater. CP, cathodic protection; FC, free corrosion; SW, sea water (data based on HSE 1999).

DNVGL-RP-C203 where N_0 was taken as 10^6 cycles for seawater with CP based on statistical assessment of the data and calibration of the *S–N* curve with the performance of existing offshore structures. An N_0 value of 10^8 cycles is used in API RP 2A which is used largely in the Gulf of Mexico where structures are not as susceptible to fatigue damage due to the different environmental criteria. The effect of the larger N_0 value is to reduce the predicted fatigue life. There is no change in slope in the *S–N* curve for free corrosion and this highlights the increased likelihood of accelerated fatigue damage in older structures if corrosion protection systems are not maintained adequately. An example of a simplified fatigue calculation taking into account a changing corrosive environment during the life is included in Appendix C.

4.5.4.3 Low Cycle/High Stress Fatigue

The low cycle/high stress range of the *S–N* curve is defined as $N < 10\,000$ cycles. The *S–N* method of fatigue life assessment is applicable to cyclic stresses in the elastic range. However, in offshore installations the presence of stress concentrations at nodes and other connections can, in some circumstances, lead to cyclic stresses that exceed the yield stress of the material locally, particularly in the low cycle region which is characterised by strain rather than by stress. It is usual to extrapolate the *S–N* curve to the low cycle fatigue region in the assessment of offshore structures due to the absence of data (as a result of the practical difficulty of conducting relevant tests which require large amplitudes) but a review of low cycle/high stress fatigue (HSE 2004) indicated that a fatigue life less than that predicted by accepted design curves is possible in the low cycle/high stress region as was observed in laboratory tests. The very high levels of cyclic stress applicable to such situations would not prove acceptable in static strength checks and would not be expected to occur in actual offshore installations due to yielding of the material. However, earlier generations of offshore structures were designed to static strength design criteria that did not reflect current knowledge and, significantly, did not address the level of extreme loading included in design practice, namely the 100-year storm event and, more recently, the 10 000-year storm event. It is recommended in DNVGL-RP-C203 that a low cycle fatigue assessment should be performed when non-linear analysis is used to determine the integrity of an offshore structure for an extreme storm event and further guidance is provided in NORSOK N-006.

Low cycle/high stress fatigue is particularly relevant to piles. Historically, there has been a limited amount of analysis to consider the effect of cyclic loading on pile performance. However, since the 1980s it has been common to perform a full fatigue analysis of piles, taking into account the effect of significant cyclic loading during pile driving (including shock wave reflection) and the effect of wave loading during operation. Low cycle/high stress fatigue failure has also been experienced during transport of jacket structures to field, due to the cyclic loading from motion and acceleration from the barge. In addition, low cycle/high stress fatigue is relevant for transverse and longitudinal bulkheads for ship-shaped structures with full load reversals and loading–unloading cycles during the life (DNVGL-RP-C206).

The pile driving process introduces very high stress range fatigue cycling and can therefore be expected to be very damaging to the integrity of the piles. This affects the remaining life and therefore is an important consideration in the management of life extension. The stresses induced by pile driving are both significant and compressive. The magnitude of the stress can be as high as the compression yield strength of the material with the potential to cause yielding of the material. The driving of the pile into the seabed results in the amplification of the stress range by reflection of a shock wave which results in a high tensile stress of a magnitude that is comparable with the compression load. Thus, the tensile part of the cyclic stress may be as high as the tensile yield stress of the material and even cause yielding to occur.

A factor which may influence the effective stress range for high cycle as well as low cycle fatigue, mean stress and stress level is the level of residual stresses in the weld. High tensile residual stresses are assumed to exist in pile welds and these are usually taken to be of yield strength in magnitude. The superposition of high tensile residual stresses with the applied stresses can have significant effects: a positive effective mean stress will be introduced, the compression element of the stress cycle will be reduced in magnitude and the tensile element will be increased, possibly leading to plasticity. Compressive cyclic stresses are considered to be less damaging than tensile cyclic stresses. For this reason, it is common to post-weld heat treat piles after fabrication to reduce the level of residual stresses. This lowers the effective mean stress, both during piling and in operation, and causes a larger proportion of the stress cycle to be in compression. This reduces the tensile stress cycle and therefore reduces the fatigue damage.

Conservative assumptions about the residual stresses are necessary because of the inaccessibility of piles for inspection. It should be noted that high residual stresses may be reduced by the pile driving process as a result of the interaction with the applied loads from the piling process as well as by the subsequent cyclic fatigue loading experienced in service, a phenomenon often referred to as 'shakedown'. This would suggest that, whilst significant damage is possible during piling, the damage in service may be lower than expected assuming the conservative assumptions made about pile fatigue – this is a consideration to be taken into account when reviewing the integrity of ageing offshore structures.

However, there is very limited information on the effect of pile driving and long-term operation on the residual stresses and, indeed, the fatigue damage to piles during the service life. This adds to the uncertainty of managing the structural integrity of offshore structures operated beyond the original design life. A study on both the fatigue capacity and the residual stresses in piles from a decommissioned offshore structure with a service life of approximately 30 years (the Edda platform) is reported in Lotsberg et al. (2010). Examination of the welds found no evidence of fatigue cracking. The results are very limited but it was noted that some of the residual stresses were of yield strength magnitude, suggesting little shakedown. The fatigue test results supported the use of the D $S-N$ curve or the E $S-N$ curve in DNVGL-RP-C203 (DNVGL 2016a) in conjunction with the misalignment SCF for butt welds. Lotsberg (2016) presents a reliability study on the pile driving and operational phases of the fatigue life of a pile of 100 mm wall thickness. Lotsberg concludes that a design factor on fatigue life of 3, which compares with a factor of 10 in ISO 19902 for uninspectable critical locations, is appropriate for piles in life-extended platforms provided that the pile driving stress cycles have been controlled and documented.

Considerable implications are likely to follow regarding the status of existing foundations if the ageing processes cannot be understood sufficiently well to be considered confidently in reassessment studies.

4.6 Fracture Mechanics Assessment

4.6.1 Introduction

Fracture mechanics analysis provides a complementary approach to the $S-N$ fatigue life assessment of offshore structures and has a particularly useful role for the assessment of ageing installations and life extension.

Unlike the conventional $S-N$ approach, it enables the assessment of defects detected during fabrication and in-service inspection. In principle, it provides a more detailed method of predicting remaining life than the $S-N$ approach. A distinct advantage is that it enables assessment for life extension by analysis of the parameters relevant to the life extension phase – it has the flexibility to take account of the precise geometry and changes in the loading and it enables both the severity of detected defects and the remaining life to be evaluated. Furthermore, using deterministic and probabilistic approaches, it provides a means of scheduling the extent and frequency of inspections and determining appropriate inspection techniques based on the accepted level of risk. However, fracture mechanics analysis is founded on the assumption that a defect is present. Where defects are not detected, a defect size corresponding to the limiting size detectable by the applied NDT method is normally assumed and relies on specialist knowledge of the interpretation of NDT data and the selection of suitable dimensions associated with probabilities of detection and sizing to ensure safety but also to avoid introducing excessive conservatism. Furthermore, it should be noted that the fracture mechanics method assumes that general linear elastic analysis principles apply beyond the local yielding that occurs in the immediate vicinity of a crack or defect and is not applicable to very small, i.e. microscopic, defects where non-linear effects dominate the region of the crack, particularly at weld joints with high stress concentrations.

Defect assessment procedures for offshore structures have been developed over the years for the design, fabrication and inspection of welds. These were very limited when the first structures were installed. However, a considerable amount of research has been performed over the years and this has led to a better understanding of the causes of structural failure and subsequently to improved guidance on integrity assessment. The principal standards applied to offshore structures are BS7910:2015 (BSI 2015) and API 579 (API 2016).

Particular applications are:

- Assessment of the remaining fatigue life of a joint in which fatigue cracks already exist.
- Assessment of the need to undertake a repair.
- Determination of the optimal frequency of in-service inspection.
- Assessment of the requirement for post-weld heat treatment during fabrication or after weld repair.
- Assessment of the effects of variations in geometrical or stress parameters for a given detail.
- Assessment of joint details that are not adequately represented by simple joint classifications.
- Assessment of the structural integrity of castings, which are used in some offshore structures.

The BS 7910 procedure is applicable to a very wide range of geometries. The assessment of plated offshore structures is covered by the wider guidance on plated welded joints. BS 7910 includes guidance specific to offshore tubular structures that can be applied in design and during in-service inspection. Special attention is focused on the assessment of known or assumed weld toe flaws, including fatigue cracks found in service in brace or chord members of T, Y, or K joints between circular section tubes under axial and/or bending loads. The guidance is also suitable for the assessment of other details which may be found in offshore structures, such as girth welds in tubular members, attachments to tubes and welds in plate structures. BS 7910 includes a probabilistic assessment procedure, see Section 4.7.

The basic components of the fatigue crack growth and fracture assessment procedure for welded joints in offshore structures are:

1. *Global structural analysis.* Determination of components of brace nominal stress corresponding to fatigue and storm loads generated by wave loading.
2. *Local joint stress analysis.* Determination of the hot-spot SCFs and the degree of bending, i.e. the proportion of the bending to total stress through the wall thickness, relevant to the crack location.
3. *Determination of stress spectrum.* Generation of the hot-spot stress range histogram for the joint.
4. *Fatigue crack growth analysis.* Integration of the appropriate fatigue crack growth law to determine the remaining fatigue life.

5. *Fracture analysis.* Use of the Failure Assessment Diagram (FAD) to determine the defect size at which fracture or plastic collapse would be expected to occur, as discussed below.

The above principles are relevant to both plated and tubular offshore structures, though specific approaches are required for the static strength/plastic collapse assessment to reflect the static strength analysis methods for the different types of structure.

4.6.2 Fatigue Crack Growth Analysis

Fatigue crack growth predictions using fracture mechanics are based on the use of a fatigue crack growth law. Paris and Erdogan (1963) established that the rate of fatigue crack growth was related to the range of the stress intensity factor, ΔK, as shown in Figure 4.10. There are three stages of crack growth:

- Stage I: Crack propagation is considered to occur only when the stress intensity factor range exceeds the threshold stress intensity factor range, ΔK_{th}.
- Stage II: At intermediate values of ΔK, there is an approximate linear relationship between the crack growth rate and ΔK on a log–log scale. This is generally characterised by the Paris equation:

$$\frac{da}{dN} = C(\Delta K)^m$$

where C and m are material constants.
- Stage III: This is characterised by accelerated crack growth which becomes unstable and results in fracture when the maximum stress intensity factor attains a critical level, K_{cr}.

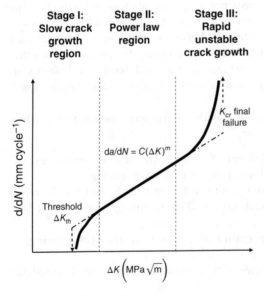

Figure 4.10 Fatigue crack growth rate curve.

Table 4.5 Fracture mechanics life assessment.

Loading data	Material data
• Response analysis • Load–time histories • Stress analysis • Cycle counting • Stress spectrum	• Material parameters • Environment for selection of crack growth curve • Crack growth curves (e.g. BS7910)

The Paris law is applicable to the Stage II region only. Various other fatigue crack growth laws have been proposed to take account of Stage I crack growth, which can represent a significant proportion of the total fatigue life, and other factors, such as the R-ratio to account for mean stress effects. A review of fatigue crack growth laws is provided in Bathias and Pineau (2010).

The fatigue life or remaining fatigue life is determined by the integration of the fatigue crack growth law. The key inputs to a fatigue crack growth assessment are summarised in Table 4.5.

The calculation of fatigue life entails integrating the Paris law:

$$N = \int_{a_i}^{a_f} \frac{da}{C \cdot (\Delta K)^m}$$

where a_i is the initial crack size and a_f is the final or critical crack size.

The integration of the fatigue crack growth law provides a means of relating the number of fatigue cycles, defect dimensions, component geometry, the applied loading and the material properties to enable predictions of remaining fatigue life or critical crack size and therefore is particularly relevant to the assessment of ageing and life extension.

A detailed reassessment of fatigue crack growth rate data for offshore structural steels was performed in King et al. (1996). The review included:

- Medium and higher strength steels with yield strengths up to 1000 MPa in air and in seawater under freely corroding conditions and under CP at levels of −850 and −1050 mV Ag/AgCl.
- Data for austenitic stainless steels in an air environment.
- Threshold stress intensity factors for both ferritic and austenitic steels.
- The effect of R-ratio due to mean and residual stress effects.
- Benchmarking the recommendations of the study against experimental S–N data for welded joints.

Least squares curve fits were applied to the data to obtain the Paris constants C and m and the standard deviation so that the plus two standard deviation design curves could be determined. The mean crack growth rates in seawater with free corrosion were found to be faster than those in air by a factor which varied between 2.4 and 3.0, depending on the level of mean stress, with no evidence of a threshold at low values of ΔK. This is consistent with the recommendation in the revised HSE fatigue guidance (incorporated in ISO 2007) of an environmental reduction factor of 3 on the welded plate and tubular joint T' S–N curves and no change in slope at 10^7 cycles for free corrosion.

The fatigue crack growth relationship is more complex for seawater with CP since hydrogen charging of the steel in the crack tip region can promote other mechanisms. This has the effect of causing an increase in crack growth rates at higher values of ΔK and this is reflected in the S–N curves which show reduced fatigue life in seawater with CP. Bi-linear curves are recommended for fatigue crack growth in air, seawater with CP and in seawater with free corrosion. Data are given for the mean and mean plus two standard deviations curve fits for $R < 0.5$ and $R > 0.5$. Unless detailed information is known about the applied loading and residual stress distribution, it is recommended that the constants for the mean plus two standard deviations and $R > 0.5$ are used.

The data for air are applicable to steels (excluding austenitic steels) with yield strengths up to 800 MPa at temperatures below 100 °C.

The recommendations for seawater are based on fatigue crack growth rate data obtained in artificial seawater or 3% NaCl solution at temperatures in the range 5–20 °C and frequencies of 0.17–0.5 Hz (typical of offshore conditions) and are consequently limited to these ranges. They are valid for steels (excluding austenitic steels) with yield strengths up to 600 MPa.

The threshold stress intensity factor range, ΔK_{th}, is related to the R-ratio for fatigue crack growth in air and in seawater with CP by the following equation for $0 < R < 0.5$:

$$\Delta K_{th} = (170 - 214 \cdot R) \, \text{N mm}^{-2/3}$$

giving values for ΔK_{th} of 170 and 63 N mm$^{-3/2}$ at R-ratios of 0 and 0.5, respectively. A threshold stress intensity factor range of zero is recommended in BS7910 for seawater with free corrosion. The fatigue crack growth rate characteristics reflect the trends with respect to the change in slope in the S–N curves for seawater and free corrosion.

The stress intensity factor range is given by the general expression

$$\Delta K = K_{max} - K_{min}$$

where

$$K = Y \cdot \sigma \cdot \sqrt{\pi \cdot a},$$

where σ is the stress and a is the crack depth and

$$\Delta K = Y \cdot \Delta\sigma \cdot \sqrt{\pi \cdot a}$$

Stress intensity factor solutions for cracks in plates are readily available. However, solutions for tubular joints are more limited. Whilst numerical methods provide the most realistic predictions of stress intensity factors, the determination of stress intensity factor solutions for cracks in tubular joints by numerical methods requires complex modelling and stress analysis and consequently only a limited number of solutions are available, e.g. Rhee et al. (1991) and Ho and Zwernemann (1995). The most extensive solutions are those obtained from finite element analysis performed on T joints and Y joints containing a semi-elliptical surface crack at the saddle point in the chord. Solutions are given for the deepest and surface points of the crack. A subsequent study by Bowness and Lee (2002) developed stress intensity factor solutions for tubular T joints from solutions for semi-elliptical surface cracks in T-butt joints.

The weight function method provides an alternative approach to the evaluation of stress intensity factor solutions and is useful for complex geometries such as tubular joints. Examples of weight functions are given in Niu and Glinka (1987) and Shen and Glinka (1991).

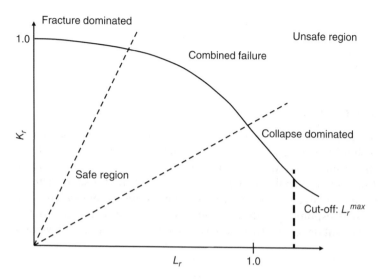

Figure 4.11 Failure Assessment Diagram (FAD).

4.6.3 Fracture Assessment

The fracture assessment procedure is based on the use of the FAD, see Figure 4.11, which combines considerations of fracture with plastic collapse. The fracture parameter, K_r, is expressed in terms of the plastic collapse parameter L_r, i.e. $K_r = f(L_r)$. K_r is derived from the stress intensity factor components for the primary and secondary stresses normalised to the material fracture toughness and L_r is the ratio of the primary load to the plastic collapse load. The fracture toughness is determined in accordance with BS EN ISO 15653:2018. The acceptability of a flaw is based on determining whether the assessment point lies within the boundary of the FAD which is represented by the function $K_r = f(L_r)$ and a cut-off point based on the maximum value of L_r^{max}, beyond which plastic collapse would be predicted to occur:

$$L_r^{max} = \frac{(\sigma_y + \sigma_u)}{2 \cdot \sigma_y}$$

L_r^{max} is derived using mean material property values.

Three levels of assessment are available: Option 1, Option 2, and Option 3 in order of increasing complexity and decreasing conservatism. Option 1 FAD does not require the stress–strain properties of the material to be known. Option 2 FAD requires the mean stress–strain curve up to σ_u to be known. Option 3 FAD is specific to the material properties, geometry and loading of the component being assessed and entails the evaluation of the elastic and corresponding elastic-plastic J-integral for the relevant loads, to yield a plot of K_r versus L_r.

The fracture assessment procedure for offshore structures is usually based on the use of the Option 2 FAD for low work hardening materials which represents the behaviour of offshore structural steels. All loading effects should be determined in the first instance for best estimates of the maximum loading, excluding load factors prescribed by limit state design codes to avoid unnecessary conservatism. The BS 7910 procedure allows the collapse parameter L_r for tubular joints to be calculated using either local collapse or global collapse analysis. The local collapse approach, which is applied to the structural component, is usually very conservative whilst the use of the global approach, which incorporates more realistic boundary conditions, tends to give more realistic predictions of plastic collapse in tubular joints.

4.6.4 Fracture Toughness Data

The crack tip opening displacement (CTOD), δ_c, is the fracture toughness parameter generally used for offshore structures, though the K_{Ic} fracture toughness parameter is used for pipelines. The value of fracture toughness used in the assessment should be based on a statistical analysis. This analysis should be based on data for the same material and fracture mechanism. The equivalent fracture toughness value is considered to be the characteristic value corresponding to the 20th percentile with 50% confidence or the 33rd percentile with 50% confidence and treated as a mean minus one standard deviation value.

4.6.5 Residual Stress Distribution

Residual stresses are an important consideration in the fracture assessment of welded joints. PD 6493:1991, recommended that the as-welded residual stress distribution should be assumed to be a pure membrane stress equal to the yield strength of the material at room temperature.

However, this guidance was generally found to be very conservative when applied to offshore structures and guidance on upper bound through-thickness residual stress distributions in a range of welded joint geometries relevant to offshore structures, namely nodal joints, circumferential pipe butt welds, pipe seam welds and T-butt welds, and repair welds were incorporated into BS 7910.

In structures that have been post-weld heat treated in an enclosed furnace in accordance with BS 5500, BS7910 recommends that the axial residual stress component, Q_m, should be assumed to be 30% of the yield strength of the material in which the defect is located for stresses parallel to the weld and 20% of the lesser of the yield strengths of the weld and parent material for stresses transverse to the weld. It should be noted that higher levels of residual stress may be present with local heat treatments at specific welds (as opposed to heat treatments applied to more extensive amounts of the structure) to code requirements and specific assessments should be made for each case.

4.6.6 Application of Fracture Mechanics to Life Extension

Fracture mechanics analysis is a particularly useful tool for assessing the integrity of a structure beyond its design life and for assessing the significance of defects and damage which are normally associated with ageing and life extension. Specific applications include:

- The prediction of the remaining fatigue life.
- Assessment of the criticality of a defect/the need for repair.
- Assessment of structural modifications or changes in the loading.
- Inspection planning and optimisation of inspection intervals.

The key strength of fracture mechanics is that it relates the relevant parameters, i.e. defect dimensions, structural geometry, loading, failure load and remaining fatigue life, through the stress intensity factor (K), the fatigue crack growth law and the fracture toughness. Thus, the method is sufficiently flexible to allow for variations in the design assumptions and account to be taken of defects to quantify the integrity of ageing structure and the extent of any life extension phase. This is in contrast to the $S–N$ method which, based on the use of the stress cycle spectrum in conjunction with design $S–N$ curves, enables the prediction of the design life and does not allow account to be taken of detected defects.

Calculation of the remaining life of offshore structures is based on the assumption that cracks initiate from defects and propagate under environmental fatigue loading. The fracture mechanics prediction is sensitive to the input parameters, particularly to the applied loading and the defect dimensions – careful consideration of these is required if meaningful results are to be obtained. This requires inspection of the structure using suitable techniques to establish the relevant defect dimensions. Where defects are not detected, the limits of the inspection technique are applied. The complex geometry of offshore structural components introduces uncertainties in the evaluation of the stress intensity factor. There is a limited range of K solutions for tubular joints and standard plate solutions for two-dimensional cracks are used in most cases. The application of plate solutions tends to yield conservative predictions of remaining life.

Defects are a feature of ageing and fracture mechanics analysis enables the assessment of the criticality of these defects. Assessment of the remaining life enables decisions to be made on the need for defect repair and evaluation of mitigation measures, e.g. defect removal by grinding/weld profiling. Fracture mechanics also provides a means of predicting the remaining life of joints containing through-thickness defects. With much reliance on the use of flooded member detection (FMD) for inspection of the substructure, fracture mechanics can provide additional information on the criticality of a defect, supplementing information on the robustness of the structure, and enables repairs to be scheduled appropriately. Fracture mechanics analysis of tubular joints containing through-thickness defects indicated that the remaining life from penetration of the wall thickness to member failure is short, with predicted periods of 0.5–3 years.

Other applications of fracture mechanics to ageing/life extension include the assessment of structural modifications by modelling the modified structure and reassessment of the remaining life using the updated applied loading which may be due to updated metocean data, structural reanalysis or revised design code criteria.

Fracture mechanics analysis is particularly relevant to the management of ageing and life extension by inspection. Assessment of defects enables inspection frequencies to be evaluated from predictions of crack growth which yield information on the remaining fatigue life. The aim of the inspection process is to detect defects before they propagate to a size which will cause structural failure and to undertake repair of the component. This has led to the development of probabilistic inspection methods which are aimed at the identification and inspection of the welds which are most critical to the integrity of the structure. An example of a fracture mechanics calculation for a semi-submersible being evaluated for life extension with different initial crack sizes is included in Appendix C.

4.7 Probabilistic Strength, Fatigue, and Fracture Mechanics

It appears that the probabilistic analyst in actuality says: 'Give me the probability densities of random variables or random functions involved and I will calculate the reliability of the structure!'. This reminds us of the well-known statement by Archimedes: 'Give me a firm spot on which to stand, and I will move the earth!'.

Elishakoff (2004)

4.7.1 Introduction

The prediction of strength and fatigue crack growth requires the use of data which are often subject to considerable uncertainty. As described in Section 2.2, this uncertainty is dealt with in design methods by using characteristic values that should be applicable for the wide range of structures and materials that the standards are intended to cover. To ensure that these standardised characteristic values are sufficiently safe for all possible structures that could be designed according to a standard, design values may be chosen to be on the safe side. For a specific structure, the standardised values may be sufficiently accurate in life extension.

The calculation of strength and fatigue life using standard approaches is subject to statistical variation and uncertainty associated with three aspects of the modelling process:

- the marine environment, response, and slowly varying loads.
- the structure.
- the capacity.

In addition, uncertainties can also be introduced during the fabrication process (misalignments, welding defects, etc.). For ageing structures, uncertainties will also be influenced by degradation mechanisms and new knowledge (improved design codes, increased knowledge of the structural behaviour, etc.).

An alternative to the standardised design approach is to use probabilistic methods, i.e. SRA, to allow for these uncertainties and to determine the probability of limit state failure and crack development.

Probabilistic methods can also be used to plan inspection intervals and to simultaneously update the reliability of the structure after inspection or repair, which is discussed further in Chapter 5.

The successful application of probabilistic procedures requires a high level of expertise and experience and probabilistic assessments should be undertaken only by appropriate specialists. Furthermore, limit state failure predictions are very sensitive to the input data. Unfortunately, adequate data may not always be available and extreme care should therefore be exercised before assumptions and approximations are made.

Establishing adequate probabilistic models for the variables is in most cases the major challenge of the reliability analysis. The probability models and their parameters should be a good representation of real data. Determination of the probability distributions is further discussed in Haldar and Mahadevan (1999). In practice, sufficient information may not be available for all the critical variables, e.g. the initial defect size and the material parameters. It is recommended in such situations that the uncertainty in the available information on these variables is included in the probabilistic models used, based on expert advice, and a sensitivity analysis performed.

4.7.2   Structural Reliability Analysis – Overview

SRA is used to analyse limit state failure and associated probabilities of load–strength systems. An overview of SRA can be found in e.g. Ang and Tang (1975, 1984), Bury (1975), Toft-Christensen and Baker (1982), Madsen et al. (1986) and Melchers (1999).

The performance of a component is described by a limit state function g. The limit state function is a function of a set of random variables $\mathbf{X} = (X_1, X_2, \ldots, X_n)$ describing the load and capacity of the structural component. Properly formulated, the event $g(\mathbf{X}) \leq 0$ defines the limit state failure of the component. Hence, the limit state of a system may be written as a function of variables $X_1, X_2 \ldots, X_n$ such that:

$$g(X_1, X_2 \ldots, X_n) = \begin{cases} > 0 \; safe \; state \\ = 0 \; limit \; state \\ < 0 \; failure \; state \end{cases}$$

where $g(\mathbf{X}) = 0$ is known as a limit state surface and each X indicates the basic load or resistance variable.

The limit state functions $g(X_1, X_2 \ldots, X_n)$ used in SRA can be expressed as shown in Table 4.6.

The probability of this limit state failure of the component is then given by the probability:

$$P_f = P[g(\mathbf{X} \leq 0)]$$

The reference period corresponding to the calculated failure probability is defined by the reference period for load in the limit state function for structures with strength that is constant (independent of time) and where the load is taken as the maximum load in a given reference period. This reference period would typically be one year or the design life of the structure giving an annual failure probability or a life failure probability.

Table 4.6 Limit state functions used in structural reliability analyses.

Strength analysis	$g = R - S$
	where R is a random variable describing the uncertainty in the strength of the structure and S is a random variable describing the uncertainty in the loading on the structure
S–N fatigue	$g = \Delta - D$
	where Δ is a random variable describing the uncertainty with the fatigue accumulation (normally with a mean of 1.0) and D is the accumulated damage calculated by the Miner summation
Fracture mechanics	$g = a_c - \delta a$
	where a_c is a random variable describing the uncertainty with the critical crack size and δa is describing the crack growth with time
	Alternatively, the crack growth can be described by the number of cycles as:
	$g = N_c - N$
	where N_c is a random variable describing the critical number of cycles defined by fracture mechanics and N is a random variable describing the number of cycles experienced

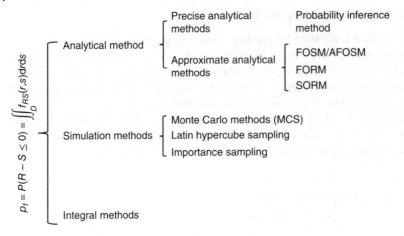

Figure 4.12 Probability of limit state failure calculation methods. MCS, Monte Carlo simulation.

In the simplest form, two variables would be applied, representing the strength, X_1, of the component and the load, X_2, on the component. The limit state function would then take the form $g(\mathbf{X}) = X_1 - X_2$. The difference $Y = g(\mathbf{X}) = X_1 - X_2$ is called the safety margin of the component.

If X is described by a joint probability density function f_X, the limit state failure probability of a structural component with respect to a single failure mode can formally be written as:

$$P_f = \int_{g(\mathbf{x}) \leq 0} f_X(\mathbf{x}) \cdot d\mathbf{x}$$

In general, this integral cannot be solved analytically. Numerical methods, simulation methods such as Monte Carlo simulations or semi-analytical approximate methods, such as FOSM (First Order Second Moment), AFOSM (Advanced First Order Second Moment), FORM (First Order Reliability Method) or SORM (Second Order Reliability Method), may be used, see overview of methods in Figure 4.12. Different levels of reliability analysis are possible depending on the level of detail applied in the uncertainty modelling. The Level I approach is based on one (characteristic) value for each uncertain parameter and is the basis for the partial factors method discussed in Section 2.2.2.2. The Level II method (FOSM and AFOSM methods) incorporates two values (mean and standard deviation) for each uncertain parameter, in addition to including correlation between parameters. The Level III reliability analysis method (FORM, SORM, and Monte Carlo simulation methods) includes a joint probability distribution function for all involved uncertain parameters and is the commonly used method at present.

4.7.3 Decision Making Based on Structural Reliability Analysis

The 'probabilities of failure' which can be calculated or estimated in civil engineering have no real statistical significance: rather, they are conventional, comparative values. Provided this point is clearly understood and accepted, probabilistic method can play a very important role in making rational comparison possible between alternative structural designs. Otherwise, they are vulnerable to all sorts of criticism

Augusti et al. (1984)

As noted in Section 4.2, some standards indicate that SRA may be used to assess existing structures for life extension (e.g. ISO 2394, ISO 13822, ISO 19902). In general, they indicate that there are limitations in the use of SRA, e.g. related to the knowledge and skill of the analyst and the data upon which the analysis is based. It is typically recommended in these standards that thorough validation of the techniques and their application is undertaken and that acceptance criteria are agreed between the regulator and the owner. For these reasons, it is not possible at present to provide acceptance criteria in ISO 19902 for the use of SRA, which indicates the difficulty in using this approach in practice.

Normally, the calculated probability of limit state failures is seen as notional or knowledge based (Moan 1997; Aven 2003) and cannot be directly compared with objective acceptance criteria such as those based on society's acceptance of structural failure. There are good reasons for evaluating the probabilities that are calculated by SRA not to be estimates of 'true' objective values but rather as assigned knowledge based probabilities (or as notional probabilities). Hence, the use of simple acceptance criteria for probabilities of limit state failure (e.g. in the form of $P_{limit\,state\,failure} \leq P_{acceptable}$) should be used with care. First, the calculated probabilities are not true values and cannot be compared with values such as objective acceptance criteria. Secondly, the views of stakeholders such as regulators, workers and society need to be heard in the decision-making process. Decisions may be based on an ALARP methodology taking into account the knowledge that the probabilities and consequences are based on or by a multi-attribute analysis that considers several stakeholders' interests (Ersdal 2005; Aven 2012).

Addressing these issues, Moan (1997) describes the recommended recipe for establishing target probabilities (acceptance criteria) as: 'The target safety levels should be based on the failure probability implicit in current codes. This method is advocated for component design checks of new structures to ensure consistency with existing design practice and the reliability methodology applied. It is emphasized that the failure probabilities which are compared to this target level should be calculated by the same methodology as used to establish the target probability. Moreover, the uncertainty measures, reliability method and target level together with the relevant procedures for load and resistance assessment should be consistent with acceptable design practice. Caution should be exercised in using general target values, i.e. without the justification of consistency as mentioned above. This is because it has been clearly demonstrated that the difference in uncertainty measures and reliability methodology used for offshore structures (jackets), varies to an extent that may yield widely scattered prediction of failure probabilities'.

An alternative to calibrating the target safety levels is proposed in Moan (1997), suggesting the following process:

- Selection of the reliability methodology (distribution functions for the parameters, model uncertainties, etc.).
- Establishment of target probabilities based on requirements in current codes.
- Assessment of the structure to be evaluated using the same reliability methodology according to this target probability.

An alternative approach is to use the Moan (1997) procedure with calibration of the target probability to existing structures that are regarded as acceptably safe instead of calibrating the target probabilities based on requirements in current codes.

4.7.4 Assessment of Existing Structures by Structural Reliability Analysis

Probabilistic methods are in many ways the most relevant method for assessing existing structures for life extension, as they enable direct evaluation of the safety of the structure taking into account all types of uncertainties, including the changes in uncertainty about the structure as it gets older. Such information is incorporated in the SRA by means of updating (Madsen et al. 1986; DNVGL 2015a). However, existing codes and regulations are not sufficiently developed with respect to making decisions based on a probabilistic approach. The purpose of SRA should be to support the decision making, not making the decision, as the calculated probabilities are an indication of a knowledge based probability of limit state failure rather than an estimate of a 'true' value of the probability of limit state failure. However, the method is very useful if used correctly and when compared with the probability of limit state failure implicit in standards (using the same probabilistic models).

The application of SRA to life extension enables the inclusion of all uncertain parameters in the strength and fatigue analysis. In addition, the degradation and the uncertainty around future degradation can also be modelled. However, there is a lack of standardised methods to include such information. Probabilistic fatigue analysis of existing structures is discussed to some extent in DNVGL-RP-C210 (DNVGL 2015a).

SRA may provide an alternative assessment method for structures that fail to satisfy the partial safety factor methods in standards which may also introduce greater accuracy in the reliability prediction. However, as indicated earlier, the decision-making has to be performed with care.

A SRA is not always necessarily a very difficult task. As an example, the limit state function for fatigue reliability, i.e. $g = \Delta - D$, in the simplest form of closed form damage calculation is:

$$g = \Delta - \frac{n}{A} \cdot q^m \cdot \Gamma(1 + {}^m\!/_h)$$

where q and h are the parameters of a Weibull distribution function representing the long-term stress range distribution and A and m are the parameters of the S–N curve. This limit state failure function could also be expressed using S_{max} (the maximum stress range in the long-term stress range distribution) directly:

$$g = \Delta - \frac{n}{A} \cdot \frac{S_{max}{}^m}{\ln(n)^{m/h}} \cdot \Gamma(1 + {}^m\!/_h)$$

Such relatively simple limit state problems can be solved by simulation or by simple FORM iteration schemes in mathematical programs such as MathCad or in spreadsheets.

The two main applications of SRA for fracture mechanics analysis in life extension are:

- Probabilistic evaluation of crack growth from a known crack, in order to include uncertainties related to the crack size, the loading and the material parameters, etc..
- Probabilistic inspection planning (see Section 5.2 for more on inspection planning).

The basic elements of a probabilistic fatigue crack growth and fracture reliability assessment are given in BS 7910. These include:

- Specification of the probability distributions and parameters for the relevant parameters, such as the initial flaw size, the final flaw size, the fatigue crack growth rate parameters (including the threshold stress intensity factor range), the stress intensity factor and the stress ranges.
- Specification of the fatigue crack growth law, such as Paris law or suitable alternatives.
- Specification of the method to be used to calculate the stress intensity factor and the failure criterion.
- Definition of the requirement, e.g. determination of the probability of failure or the inspection interval, and establishment of the corresponding limit state function.

Probabilistic fatigue crack growth and fracture assessment are dependent on the availability of a significant amount of information on parameters and their distributions. A particularly large source of uncertainty in fracture mechanics assessment is the information about the defect size and distribution. In addition, there are uncertainties about representation of the $\log(da/dN)$ versus $\log(\Delta K)$ plot, modelling of crack growth in complex joints and the calculation of the fatigue loading. Further information about the probabilistic modelling of these parameters can be found in BS 7910 and DNVGL-RP-C210.

References

ABS (2014). ABS Guide for Fatigue Assessment of Offshore Structures. American Bureau of Shipping (ABS).

Ang, A.H.S. and Tang, W.H. (1975). *Probability Concepts in Engineering Planning and Design, Volume I – Basic Principles*. New York, NY: Wiley.

Ang, A.H.S. and Tang, W.H. (1984). *Probability Concepts in Engineering Planning and Design, Volume II – Decision, Risk, and Reliability*. New York, NY: Wiley.

API (2014). API RP-2A *recommended practice for planning, design and constructing fixed offshore platforms*. In: *API Recommended Practice 2A*, 22e. American Petroleum Institute.

API (2016). *API RP 579-1/ASME FFS-1, Fitness-For-Service*, 3e. American Petroleum Institute.

Augusti, G., Baratta, A., and Casciati, F. (1984). *Probabilistic Methods in Structural Engineering*. London: Chapman and Hall Ltd.

Aven, T. (2003). *Foundation of Risk Analysis, a Knowledge and Decision-Oriented Perspective*. Chichester: Wiley.

Aven, T. (2012). On the meaning and use of the risk appetite concept. *Risk Analysis: An International Journal* 33 (3): 349–504.

Bathias, C. and Pineau, A. (eds.) (2010). *Fatigue of Materials and Structures: Fundamentals*. ISTE Ltd.

Bowness, D. and Lee, M.M.K. (2002). Fracture mechanics assessment of fatigue cracks in offshore tubular structures, Report OTR 2000/077, Health and Safety Executive.

BSI (2015). BS 7910:2013 + A1:2015, 'Guide to methods for assessing the acceptability of flaws in metallic structures', British Standards Institution.

Bury, K.V. (1975). *Statistical Models in Applied Science*. Wiley.

DNV (1999). *ULTIGUIDE – Best Practice Guideline for Use of Non-linear Analysis Methods in Documentation of Ultimate Limit States for Jacket Type Offshore Structures*. Høvik: Det Norske Veritas.

DNVGL (2015a). DNVGL-RP-C210, Probabilistic methods for planning of inspection of fatigue cracks in offshore structures, DNVGL.

DNVGL (2015b). DNVGL-CG-0172, Thickness diminution for mobile offshore units, DNVGL.

DNVGL (2016a). DNVGL-RP-C203, Fatigue design of offshore steel structures, DNVGL.

DNVGL (2016b). DNVGL-RP-C208, Determination of structural capacity by non-linear finite element analysis methods, DNVGL.

Efthymiou, M. (1988). Development of SCF formulae and generalized functions for use in fatigue analysis. Proceedings of OTJ'88, Surrey, UK.

Elishakoff, I. (2004). *Safety Factors and Reliability: Friends or Foes?* Dordrecht: Kluwer Academic.

Ellinas, C.P. and Walker, A.C. (1983). Damage on offshore tubular bracing members. Proceedings of IABSE Colloquium on ship Collision with Bridges and Offshore Structures, Copenhagen, Denmark, pp. 253–261.

Ersdal (2005). Assessment of existing structures for life extension. PhD thesis. University of Stavanger.

Haldar, A. and Mahadevan, S. (1999). *Probability, Reliability, and Statistical Methods in Engineering Design*, 1e. Wiley.

Hellan, Ø. (1995). Nonlinear pushover and cyclic analysis in ultimate limit state design and reassessment of tubular steel offshore structures. PhD thesis: Norwegian Institute of Technology, University in Trondheim, Norway

Ho, C.M. and Zwerneman, F.J. (1995). *Assessment of Simplified Methods. Report on Joint Industry Project Ffracture Mechanics Investigation of Tubular Joints, Phase Two*. Stillwater, OK: Oklahoma State University.

Hørnlund, E., Ersdal, G., Hineraker, R.H. et al. (2011). Material issues in ageing and life extension, Paper No. OMAE2011-49363, 30th International Conference on Ocean, Offshore and Arctic Engineering, Rotterdam, The Netherlands (19–24 June 2011).

HSE (1997). The Durability of Prestressing Components in Offshore Concrete Structures, Offshore Technology report OTO 97 053. HSE Information Service.

HSE (1999). Background to New Fatigue Guidance for Steel Joints and Connections in Offshore Structures, HSE report OTH 92 390.

HSE (2004). Failure Control Limited, 'Review of Low Cycle Fatigue Resistance', HSE research report 207.

ISO (2000). ISO/DIS 13822, Bases for design of structures – Assessment of existing structures. International Standardisation Organisation.

ISO (2007). ISO 19902, Petroleum and natural gas industries – Fixed steel offshore structures. International Standardisation Organisation.

ISO (2013). ISO 19900:2013, Petroleum and natural gas industries – General requirements for offshore structures. International Standardisation Organisation.

King, R.N. Stacey, A., and Sharp, J.V. (1996). A Review of Fatigue Crack Growth Rates for Offshore Steels in Air and Seawater Environment. 14th International Conference on Offshore Mechanics and Arctic Engineering (OMAE), Florence, Italy (16–20, June 1996).

Kinra, R.K. and Marshall, P.W. (1980). Fatigue analysis of the Cognac platform. *Journal of Petroleum Technology*, Paper SPE 8600.

Landet, E. and Lotsberg, I. (1992). Laboratory testing of ultimate strength of dented tubular members. *ASCE, Journal of Structural Engineering* 118 (4): 1071–1089.

Lange, H., Berge, S., Rogne, T., and Glomsaker, T. (2004). Robust material selection. Report for Petroleum Safety Authority Norway, SINTEF, Trondheim, Norway.

Lotsberg, I. (2016). *Fatigue Design of Marine Structures*. Cambridge University Press.

Lotsberg, I., Wästberg, S., Ulle, H. et al. (2010). Fatigue testing and S-N data for fatigue analysis of piles'. *Journal of Offshore Mechanics and Arctic Engineering* 32.

Lutes, L.D., Kohutek, T.L., Ellison, B.K., and Konen, K.F. (2001). Assessing the compressive strength of corroded tubular members. *Applied Ocean Research* 23: 263–268.

Madsen, H.O., Krenk, S., and Lind, N.C. (1986). *Methods for Structural Safety*. Englewood Cliffs, NJ: Prentice-Hall Inc.

Melchers, R.E. (1999). *Structural Reliability Analysis and Predictions*. Wiley.

Moan, T. 1988. The Inherent Safety of Structures Designed According to the NPD Regulations, SINTEF report STF71 F88043. Trondheim, Norway.

Moan, T. (1997). Target levels for structural reliability and risk analysis of offshore structures. In: *Risk and Reliability in Marine Technology* (ed. C. Guedes Soares). Rotterdam: A.A. Balkema.

Niu, X. and Glinka, G. (1987). The weld profile effect on stress intensity factors in weldments. *International Journal of Fracture* 35 (1): 3–20.

Ocean Structures (2009). OSL-804-R04 Ageing of Offshore Concrete Structures, Myreside, UK.

Paris, P.C. and Erdogan, F. (1963). A critical analysis of crack propagation laws. *Journal of Basic Engineering* 85: 528–533.

Poseidon (2007). POS-DK07-136-ROI Specialist support on structural integrity issues. Poseidon International Ltd, Aberdeen.

Rhee, H.C., Han, S., and Gibson, G.S. (1991). Reliability of solution method and empirical formulas of stress intensity factors for weld toe cracks of tubular joints. In: *Proceedings of the 10th Conference on Offshore Mechanics and Arctic Engineering (OMAE '91), Vol. III-B, Materials Engineering* (ed. M.M. Salama et al.), 441–452. New York, NY: The American Society of Mechanical Engineers.

Saad-Eldeen, S., Garbatov, Y., and Guedes Soares, C. (2012). Effect of corrosion degradation on ultimate strength of steel box girders. *Corrosion Engineering, Science and Technology* 47 (4).

Saad-Eldeen, S., Garbatov, Y., and Guedes Soares, C. (2013). Experimental assessment of corroded steel box-girders subjected to uniform bending. *Ships and Offshore Structures* 8 (6): 653–662.

Saad-Eldeen, S., Garbatov, Y., and Guedes Soares, C. (2015). Fast approach for ultimate strength assessment of steel box girders subjected to non-uniform corrosion degradation. *Corrosion Engineering, Science and Technology* 51 (1): 60–76.

Shen, G. and Glinka, G. (1991). Weight functions for a surface semi-elliptical crack in a finite thickness plate. *Theoretical and Applied Fracture Mechanics* 15 (3): 247–255.

Skallerud, B. and Amdahl, J. (2002). *Nonlinear Analysis of Offshore Structures*. Baldock: Research Studies Press Ltd.

Smith, C.S. (1986). Residual strength of tubulars containing combined bending and dent damage. Proceedings of the Offshore Operations Symposium, Ninth Annual Energy Sources Technology Conference and Exhibition, New Orleans, LA.

Smith, C.S., Kirkwood, W., and Swan, J.W. (1979). Buckling strength and post-collapse behaviour of tubular bracing members including damage effects. Proceedings of the 2nd International Conference on the Behaviour of Offshore Structures, BOSS 1979, London, UK.

Smith, C.S., Sommerville, W.C., and Swan, J.W. (1981). Residual strength and stiffness of damaged steel bracing members. Proceedings of the 14th Offshore Technology Conference, OTC Paper No. 3981, Houston, TX (4-7 May 1981).

Søreide, T.H. and Amdahl, J. (1986). USFOS – A computer program for ultimate strength analysis of framed offshore structures; Theory manual. Report STF71 A86049, SINTEF Structural Engineering, Trondheim, Norway.

Standard Norge (2012). NORSOK N-001, Integrity of offshore structures, 8e; September 2012. Standard Norge, Lysaker, Norway.

Standard Norge (2013). NORSOK N-004, Design of steel structures. Rev. 3. Standard Norge, Lysaker, Norway.

Standard Norge (2015). NORSOK N-006, Assessment of structural integrity for existing offshore load-bearing structures, 1st edition; March 2009. Standard Norge, Lysaker, Norway.

Taby, J. and Moan, T. (1985). Collapse and residual strength of damaged tubular members. Proceedings of the Fourth International Conference on Behaviour of Offshore Structures, Delft, the Netherlands (1–5 July).

Taby, J. and Moan, T. (1987). Ultimate behaviour of circular tubular members with large initial imperfections. Proceedings of the 1987 Annual Technical Session, Structural Stability Research Council.

Tilly, G.P. (2002). Performance and management of post-tensioned structures. In: *Proceedings of the Institute of Civil Engineers, Structures and Buildings*, vol. 152, 3–16.

Toft-Christensen, P. and Baker, M.J. (1982). *Structural Reliability Theory and Its Applications*. Berlin: Springer Verlag.

Tweed, J. H. and Freeman, J. H. (1987). Remaining Life of Defective Tubular Joints, Offshore Technology Report OTH 87 259. HMSO.

Yao, T., Taby, U. and Moan, T. (1986). Ultimate strength and post-ultimate strength behaviour of damaged tubular members in offshore structures. Proceedings of the International Symposium on Offshore Mechanics and Arctic Engineering, Tokyo, Japan.

5

Inspection and Mitigation of Ageing Structures

5.1 Introduction

Inspection is an activity that is important in maintaining the safety of structures in operation, both in reducing uncertainty about their current state and to detect any defects. As mentioned in Chapter 2, the standard structural integrity management (SIM) procedures include surveillance and inspection, evaluation, assessment (if needed) and mitigations as required processes in ensuring the integrity of structures.[1]

The meaning of inspection here is the obtaining of information on the condition and configuration of the structure by means of visual examination, NDT (NDE) methods (see Appendix B) and monitoring. Inspection is particularly important for ageing structures and for collecting the necessary data for life extension. Even if defects are found, further inspections, evaluations and analysis are usually necessary to determine the seriousness of a defect and its impact on structural integrity. As a result of this evaluation, mitigation may be required to maintain structural integrity.

Mitigation provides the opportunity to repair any degradation, damage or other changes. Mitigations include repairs, improvements and strengthening of the structure. These are discussed in more detail in Section 5.4. As most mitigations are relatively expensive and time consuming, careful analysis would be required to determine the need for and the type of mitigation to be implemented. The case for life extension will often require detailed analysis and mitigations as part of the guarantee for integrity in the extended period.

A key factor for allowing a structure to be used in life extension is that it is possible to inspect it and, hence, gather sufficient information about its condition. Such information is required in the structural integrity assessment which includes that from the fabrication stage and the operational phase. The ability to collect this data using a range of techniques is determined mainly at the design stage.

ISO 19900 (2013) notes that 'Structural integrity, serviceability throughout the intended service life and durability are not simply functions of the design calculations but are also dependent on the quality control exercised in construction, the supervision on site and the manner in which the structure is used and maintained'. Therefore, it

---

1 It should be mentioned that this book is about the primary structure, which includes the substructure and the main load-bearing part of the topside, including the flare tower and helideck. However, it should be noted that many secondary and tertiary structural components which are not important for the main load-bearing capacity may be important for overall safety as they may present a hazard in a deteriorated condition, e.g. potential dropped objects which can cause structural damage or injury to personnel.

Ageing and Life Extension of Offshore Structures: The Challenge of Managing Structural Integrity, First Edition. Gerhard Ersdal, John V. Sharp, and Alexander Stacey.
© 2019 John Wiley & Sons Ltd. Published 2019 by John Wiley & Sons Ltd.

is important that during the design and fabrication stages an approach for inspection and maintenance should be developed. ISO 19902 (ISO 2007) states that, in preparing the plan for inspectability, a realistic assessment should be made of the actual ability to achieve the intended quality of the inspection and maintenance.

Another key factor for extending the life of a structure is the possibility of repairing it if structural damage and a need for strengthening are found during inspections and assessment. Although design for repairability does not feature strongly at the design stage of an offshore installation, compared with for example the ability to inspect, it becomes an important issue when repairs are required as a result of damage or deterioration due to ageing.

5.2 Inspection

5.2.1 Introduction

During the operation of offshore structures, inspection is carried out to identify any damage and degradation (e.g. cracking), particularly in welded joints. There are a number of techniques available for inspecting a structure, which have been developed over many years. A brief summary of these is given in Appendix B.

Inspection is important in reducing uncertainty about the current state of the structure. An example of reducing uncertainty would be to use inspections to verify results from, for example, a fatigue analysis. However, there are uncertainties associated with fatigue analysis and inspection results are important in developing confidence in the actual state of the structure. The identification of defects reduces uncertainty and the resulting fatigue analysis taking account of the inspection results will give more confidence in the actual state of the structure.

It should be noted that there are uncertainties also in the inspection results, due to the reliability of the inspection. An inspection will not find cracks smaller than a certain limit, dependent on the inspection method and the conditions in which the inspection is performed. The capability of detection for various inspection methods is defined as a function of defect size and the condition and is illustrated by a probability of detection (PoD) curve. As an example, DNVGL-RP-C210 (DNVGL 2015) indicates that there is a 90% probability of detecting a 12 mm deep crack underwater by alternating current field measurement (ACFM) and magnetic particle inspection (MPI) (see Appendix B). Further, there is a 90% probability of detecting a 350–400 mm crack length by close visual inspection (CVI) under difficult conditions (underwater would typically fall into this category). These factors are dependent on the ability of the operator and the PoD curves are based on data collected from many operators.

Inspection is often used in a broader sense than just checking the structure's condition. In ISO 19901-9 (ISO 2017), inspection is defined as all survey activities with the purpose of collecting the necessary data required for evaluating the integrity of the structure. Inspection would then, in addition to surveying the actual condition of the structure, also include surveillance of configuration, loads, information, knowledge, standards, regulation and other changes that affect the safety of a structure, as described in Chapter 3.

Regular inspections are a regulatory requirement in most countries with offshore installations. Inspection is, for example, a requirement of the UK safety regulations for offshore structures. Under the Certificate of Fitness (CoF) regime (which lasted until 1992), there was a five-year renewal cycle which required both annual and five-yearly inspections for the renewal of the certificate. This was replaced by the Safety Case regime in 1992 which included the Design and Construction Regulations (DCRs), where Regulation 8 included a need to maintain 'integrity' during the life cycle by periodic inspections and any remedial work as necessary. These regulations also introduced safety critical elements (SCEs), which include the structure and topsides and are required as focus for the inspections.

In Norway, the Norwegian Petroleum Directorate (NPD) issued guidelines for the inspection of primary and secondary structures in 1976. These required an initial inspection (first year inspection) and subsequent annual inspections. In addition, there was a four-yearly condition evaluation to produce a summary of the results from inspections and potential analysis of findings from these in order to derive updates to the framework inspection programme. Similar requirements have been maintained in the Norwegian regulations, both in the 1992 update of the NPD regulation and in the NORSOK N-005 standard that replaced this regulation in 1997. However, the requirement of a four-yearly update of the long-term framework inspection programme was relaxed in NORSOK N-005 and the update of this programme was left to the operator to perform inspections when necessary.

Structural inspection tasks for a fixed steel structure include the inspection of welded connections for cracking, the integrity of the foundations, the extent of corrosion and the state of the cathodic protection (CP) system (Section 5.2.5). For floating structures, inspection of the hull for corrosion and cracks with respect to watertight integrity and assessment of the integrity of the mooring system are also key issues (Section 5.2.6). For the topside structural condition, inspection for cracks in welded connections and support members is necessary, as well as checking for corrosion which can be difficult if cladding is present (Section 5.2.7).

Competence in both planning inspections and undertaking them in the field is an important factor in ensuring integrity. ISO 19902 (ISO 2007) lists several requirements for competence in the management of the inspection, maintenance and repair (IMR) database, planning the inspection programme and its offshore execution. Certification schemes exist for personnel involved in actual inspections (e.g. CSWIP 2018). These include those for divers involved in inspection, remote operated vehicle (ROV) inspectors, underwater inspector controllers and general inspectors for offshore facilities (mainly for topside). Interestingly, to the knowledge of the authors, no certification schemes exist for the engineers responsible for planning the inspections, evaluating the findings from these inspections and eventually the overall integrity of the structure. This is concerning as these engineers are a key element in the SIM process.

5.2.2 The Inspection Process

Several international standards, e.g. ISO 2394 (ISO 1998), ISO 19901-9 (ISO 2017) and ISO 19902 (ISO 2007) specify requirements for SIM which includes a process cycle for inspections, as indicated in Figure 5.1.

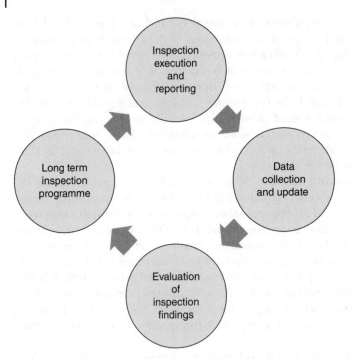

Figure 5.1 Inspection process cycle similar to that as described in ISO 19902 (2007).

The process shown in Figure 5.1 is a cycle for inspection planning, performance, reporting and evaluation, including:

- Collection and retention of data from present and previous inspections, in addition to data from design, fabrication and installation.
- Evaluation of the findings and anomalies in the data (e.g. cracks, corrosion, changes in loading, standards and knowledge, etc.).
- Update of the long-term inspection programme based on the evaluation of the data, which contains an overall plan for what needs to be inspected, when and how.
- The development of an inspection work scope which contains the detailed specification for inspection activities and the means of offshore execution and procedures for reporting data.

Several standards on SIM, e.g. ISO 19902, describe four different types of inspection of the condition of the structure:

- *Baseline*. Inspection soon after installation and commissioning to detect any defects arising from fabrication or damage during installation.
- *Periodic*. Inspection to detect deterioration or damage over time and discover any unknown defects.
- *Special*. Inspection required after repairs or to monitor known defects, damage or scour.
- *Unscheduled*. Inspection undertaken after a major environmental event such as a severe storm, hurricane, earthquake, etc., or after an accidental event such as a vessel impact or dropped objects.

For ageing structures, the periodic, special and unscheduled inspections are dominant, with expectation that the frequency of periodic inspections increases.

In addition to these international standards, several regional standards also provide relevant information about the inspection of topside, fixed and floating structures. In particular these are API RP 2SIM, the proposed API RP 2FSIM (API 2014) and API RP 2MIM (API 2018), the NORSOK N-005 (Standard Norge 2017) and N-006 (Standard Norge 2015).

5.2.3 Inspection Philosophies

Inspection of the condition of offshore structures is costly and has safety implications, particularly if divers are used. Hence, it is impractical to inspect all members, components and areas of structures. As a result, the selection of inspection methods, their deployment and the frequency of inspections are key factors in planning an effective inspection and cost-efficient programme that also provides sufficiently safe structures. Over the years there have been changes to the approach to inspection planning, as shown in Figure 5.2.

Initially, regulators and operators relied on time or calendar based inspections, as required by the CoF regime in the UK sector and the NPD regulations from 1976. These calendar based inspection plans indicated a fixed inspection interval, typically annually and up to five-year intervals depending on the application. It was, however, regarded as too prescriptive in some cases and not sufficient in other cases. These calendar based inspections found a certain number of anomalies but little damage or deterioration that had a significant impact on the integrity of these relatively new structures. This led to a condition based approach to inspections, where the interval between inspections was determined by the state of the structure identified in previous inspections. However, it was found to be difficult to prioritise inspections due to the complexity of the structures and the criticality of the welded detail was introduced into the condition based inspection planning.

As a result of the ever-lasting search for effective inspection and cost-efficient programmes, there is a need to optimise the inspection process. An alternative approach introduced in the 1990s was the risk and probabilistic based approaches where inspections were focused on areas of the structure considered to be at the highest risk of failure,

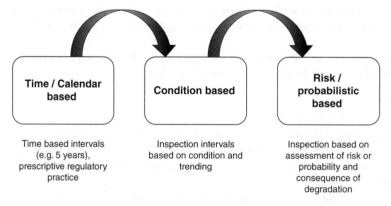

Figure 5.2 Development of planning philosophies for inspection of structural condition.

with the meaning of risk in this context as the probability and consequence of failure of a member, detail or area of a structure. An example of such an area of the structure that would be seen to be of high risk would be a welded connection with high criticality and a short fatigue life.

5.2.4 Risk and Probabilistic Based Inspection Planning

Several models have been developed for risk based prioritisation of inspections, with either a quantitative or qualitative assessment of the probability and the consequences of failure of a specific member, component or area of the structure. One of the most useful methods for a substructure is the application of structural reliability analysis (SRA) (see Section 4.7) to determine the probability of fatigue cracking in conjunction with non-linear redundancy analysis to determine the consequence of failure of this member, component or area of the structure. For jacket structures, the level of redundancy is typically described by the reserve strength ratio (RSR) and damaged strength ratio (DSR), described in Chapter 4. A few attempts to model the totality of the structural system in a SRA have been made with the aim of deriving more consistent probabilities and consequences of failure (Dalane 1993). However, this approach has proved to be rather cumbersome and has not been used much in practice.

Probabilistic methods based on SRA for planning inspections have been gaining ground in the last few decades since, in principle, they enable inspection to be optimised by targeting joints likely to accrue the most damage. The models being developed are rather complex and require input distributions for all of the main parameters. Limited data exist and the distributions are not known with any certainty. In particular, information from the fabrication inspection is not normally included. The models are usually based on optimising the reliability of the structure, as a result of detecting and repairing cracks in joints due to fatigue.

In order to establish the inspection schedules for details with calculated fatigue lives, the following probabilities need to be quantified:

- The failure probability as a function of the elapsed time.
- The probability of detecting a given crack size (PoD).
- The target probability of failure, i.e. the maximum acceptable failure probability.

The analyses are normally performed using probabilistic crack growth models. When calculating the probabilities after a planned inspection, the joint probabilities of events (the result of the inspection, the probability of detecting a crack and a future failure) need to be calculated.

The PoD is dependent on the capability of the inspection method used. For example, the PoD curve for eddy current inspection (EC) is defined by:

$$P_{Detection}(a) = 1 - \frac{1}{1 + \left(a/x_0\right)^b}$$

where a is the crack depth. The values for the parameters $x_0 = 0.161$ and $b = 1.01$ have been determined by fitting a PoD curve to experimental result for good working conditions taken from the ICON database, Dover and Rudlin (1996), and are based on 1205 observations.

An important factor is the setting of a target level of reliability (e.g. for the occurrence of fatigue cracking), which determines the level at which inspections are required. In some cases, this has been based on the level which has been achieved to date and assuming that this is acceptable for the future (despite ageing structures). In other cases, sensitivity studies have been used to explore the effect of different levels of reliability on the overall integrity. This remains one of the difficult areas in inspection planning. In published material, target reliabilities have varied from 1×10^{-4} to 2.5×10^{-2} (annual probability of fatigue crack occurrence). This is a wide range and gives rise to some concern in using this approach, highlighting the issue discussed in Section 4.7 on using target values that are calibrated to the probabilistic models used and the safety level required in the standards. For jacket structures, the effects of structural redundancy are particularly important in setting such targets. Another limiting factor is the unexpected nature of some of the failures found in practice, which will not be predicted using current probabilistic techniques.

A risk based approach would typically lead to longer intervals between inspections (compared with calendar or condition based approaches) and is often claimed to provide optimised structural integrity. However, the prediction of these longer interval contradicts the natural assumption that more inspections are needed when a structure is getting older and is expected to experience more cracks and other degradation. This indicates a concern about the use of reliability based inspection planning for ageing structures.

Disadvantages of the probabilistic SRA based inspection approach are as follows:

- It cannot predict unlikely causes of damage, such as internal cracking at welded connections (which has been found in practice but was not predicted by analysis).
- Setting the target reliability is a difficult task and the target reliability is dependent on the probabilistic models used in the SRA.
- For ageing structures, the calculated inspection intervals have a tendency to continue to increase with time, which is against expectation.

The use of risk matrices enables a more qualitative approach. A risk matrix is produced to highlight the components with the highest risk scores and hence priority for inspection. An example of a typical 5×5 risk matrix for risk based inspection planning is shown in Figure 5.3. Other configurations are possible and larger matrices are known to be used by the offshore industry to provide a more refined risk assessment.

Severity levels A–E in Figure 5.3 are typically determined based on qualitative or quantitative evaluation of the consequence of failure of the member. Severity level A typically indicates very low consequence of failure of the member, while severity level E would

Consequence				Accumulated probability				
				1	2	3	4	5
People	Environment	Cost	Severity level	$<10^{-3}$	10^{-3}–10^{-2}	10^{-2}–10^{-1}	10^{-1}–1	~1
Multiple fatalities	Massive effect	Very high	E	M	H	H	H	H
Single fatality or permanent disability	Major effect	High	D	M	M	H	H	H
Major injury	Localised effect	Medium	C	L	M	M	H	H
Slight injury	Minor effect	Low	B	L	L	M	M	H
Superficial injury	Slight effect	Negligible	A	L	L	L	M	M

Figure 5.3 Example risk matrix for risk based inspection planning. L, low risk; M, medium risk; and H, high risk.

typically indicate that the structure would collapse if the member, component or area of the structure should fail. Accumulated probability would similarly be determined either by a rigorous SRA or by assessment based on experience or databases of failure rates. However, the data are limited and this makes this approach difficult to use in practice.

In life extension, risk based inspection planning should be used with great care. A maximum interval of five years between inspections for important members is considered by some standards (Standard Norge 2015) to be good practice. For critical members and joints, a crack growth analysis based on a fracture mechanics analysis should be performed in order to determine the intervals that would ensure two inspections between the crack being detectable and becoming critical.

5.2.5 Inspection of Fixed Jacket Structures

In the early life of North Sea structures, detailed inspection of welded joints was carried out using divers. A significant amount of damage was found during inspection. Failure to detect fatigue damage has resulted in major structural failures, for example the *Alexander L. Kielland* accident in 1980 with 123 fatalities. This led to a major effort to develop suitable fatigue design and assessment methods in the 1980s and 1990s and this has resulted in a substantial reduction in the amount of fatigue damage being found. Furthermore, with greater emphasis on safety, the safety of divers operating in deep waters became an important factor, resulting in decreased use of divers and the wider use of ROV based inspection techniques.

A number of techniques have been developed over many years to enable underwater detection, from the use of divers in the early years to the current approach using ROVs. A good overview of most inspection techniques is given in NORSOK N-005 Annex B (Standard Norge 2017).

For fixed jacket structures, the current approach to underwater inspection is general visual inspection (GVI), CVI and flooded member detection (FMD). These have the key advantages of being relatively quick and hence less expensive and can cover most of the installation below water in a short period. They are normally undertaken from an ROV. However, FMD has the significant disadvantage of being able to detect only through-thickness cracks which relies on the detection of water penetrating the member. As noted in Section 4.6, the remaining life is short after detection of flooding. Hence, the efficient use of FMD relies on understanding the criticality of individual joints. There are also concerns about relying on FMD for ageing structures where cracking is more likely to occur. This may require more frequent inspections or the use of more detailed NDE methods. CVI and GVI also have limitations. As indicated earlier, the PoD of crack lengths less than 350–400 mm by CVI under difficult conditions is limited. Detecting a crack by GVI is not expected to be feasible but GVI will typically be expected to identify member severance or severe denting. When cracks have been found by CVI or FMD, it is normal to investigate further, possibly using more detailed inspection methods such as MPI.

However, since in most cases there is sufficient redundancy in jacket structures such that failure of one joint or member is unlikely to cause overall structural collapse, such inspection methods as GVI, CVI and FMD are often believed to be acceptable for redundant structures. This is because redundant fixed offshore platforms generally have a multiplicity of load paths so that failure of one component does not necessarily lead to catastrophic structural collapse. However, for ageing fixed jacket structures this may not be the case as widespread damage is expected at some stage making these inspection methods less acceptable. The topic of system strength is discussed in more detail in Section 4.4.

Inspection of the corrosion protection system (usually based on anodes attached to the structure) is undertaken by measuring local potentials (with respect to a standard electrode) using a ROV. More positive potentials may indicate high usage of the anodes, which can be assessed visually and replacement may be needed (see Section 3.4).

The inspection of guide frames and support structures for risers, conductors and caissons has historically been included within the scope of structural inspection. In the early years of oil and gas production in the Northern European region, a large amount of fatigue cracking was found in these guide frames due to inadequate design. For some of these older structures this is still a problem area requiring inspections and possibly repair but trends at present indicate that it is less of a problem for newer structures. For guide frames in or above the splash zone, corrosion will also be a major issue. These are to some extent protected by different types of coating and the integrity of these coatings need to be included in an inspection programme.

Inspection of components in the splash zone is difficult and often requires rope access or upward looking ROVs. In the last few years, the use of drones, known as aerial remote operated vehicles (AROVs), have been introduced with promising results. The ability to inspect in this zone is limited and at the design stage this is recognised by avoiding fatigue sensitive components in this area.

Underwater inspection may also include assessment for scour around foundation piles, location of build-up of seabed materials, drill cutting and debris, etc. Scour can reduce the pile efficiency as it exposes the upper parts of the piles (introducing buckling and corrosion as potential failure modes). Build-up of materials around the base of the structure reduces access to inspection of the buried members and this may also introduce unexpected loads into these members. The inspection of the degree of scour and build-up of materials around the base of the structure is normally performed by a GVI using an ROV. Removal of built-up materials maybe required if critical structural members have been buried.

Standards for the inspection of fixed structures have been developed to document the inspection process. The principal international standard for fixed jacket structures is ISO 19902 (ISO 2007). This standard includes the inspection process given in Section 5.2.2. Further, the standard both provides a calendar based inspection programme, as well as allowing a risk based inspection regime. Also, the standard specifies four levels of structural inspection:

- *Level I*. Above water visual survey, below water check on the CP system.
- *Level II*. General underwater visual survey, checking for excessive corrosion, accidental damage, scour, large cracks, etc.
- *Level III*. Underwater CVI of pre-selected areas and FMD.
- *Level IV*. Underwater NDE of selected areas, detailed CVI, etc.

The suggested calendar based frequency of inspection in ISO 19902 is annual for Level I, every three to five years for Level II, every five years for Level III and as required for Level IV. These frequencies are more likely to apply to structures in environments where fatigue is less of a problem, such as the Gulf of Mexico.

For structures in the Northern European region, the development of a more comprehensive inspection plan is normally required, describing the techniques to be used, the frequency of inspection, etc., as indicated in ISO 19902 (ISO 2007), ISO 19901-9 (ISO 2017) and NORSOK N-005 (Standard Norge 2017). However, the inspection process in these standards follows the typical work process as indicated in Section 5.2.2. Other relevant information on the inspection of fixed structures in this region is given in an Oil & Gas UK (O&GUK 2014) report on fixed structures. Similarly, inspection of fixed offshore structures in the US and many other areas of the world is performed in accordance with API RP 2SIM (API 2014). This recommended practice is similar to ISO 19902, ISO 19901-9 and NORSOK N-005 but is less focused on fatigue (as fatigue is less of a problem in US waters).

The inspection of ageing fixed structures should require a more demanding approach recognising that the ageing processes (e.g. fatigue and corrosion) are likely to occur more frequently. Hence, there will be a higher probability of damage and the potential for widespread degradation will be increasing. Any standard which includes inspection of ageing structures should take these two items into account. However, most of the existing standards do not include specific requirements for the inspection of ageing structures. This may be a significant deficiency in these standards as an increasing number of structures have reached their life extension stage and without more extensive inspection they are likely to experience more failure and multiple failures in the near future. Hence, these standards should be updated to include stronger requirements on for example inspection intervals, specific ageing mechanisms and how to deal with widespread degradation and damage in ageing structures.

NORSOK N-006 in section 9.1 (Standard Norge 2015) is the only standard that includes a number of requirements for the inspection of ageing structures. These include adjusting the inspection intervals due to:

- An increased likelihood of fatigue cracks as more fatigue damage is being accumulated.
- The consequence of failure may change as it becomes more likely that more than one joint can fail.

NORSOK N-006 is important in that it requires that inspection intervals for ageing structures should be determined such that a potential fatigue crack is detected with a large certainty prior to becoming a threat to the integrity of the structure based on fracture mechanics crack growth analysis. Components where failure can lead to substantial consequences and have passed their fatigue design life are required to be inspected by an appropriate NDT method, with the interval of inspection being based on criticality, crack growth characteristics and probability of crack detection. Furthermore, it is required that these components are inspected at least every five years, even if risk based or probabilistic calculations indicate longer intervals.

NORSOK N-006 also requires a fatigue analysis in the damaged condition for structures. This is because it is possible that more than one connection may fail due to fatigue during one winter season, which typically is when it is more difficult to repair any damage that has occurred. This implies that fatigue capacity is checked with these members in a condition where they are unable to carry load (damaged, severed). This analysis should form an input to the inspection plan to ensure that fatigue cracks are detected prior to accelerated fatigue cracking in the remaining structure due to redistribution of stresses.

HSE has published an information sheet 'Guidance on management of ageing and thorough review for ageing installations' (HSE 2009). A 'thorough review' is required as part of the revision of safety cases every five years. This guidance note identifies several factors which relate to inspection planning, particularly of SCEs, including the need to demonstrate how structural degradation can affect the performance of SCEs and a detailed knowledge of the current state of the structure. It is important to understand that this latter requirement may not be met by current inspection planning involving FMD. The Guidance note also includes a list of some aspects of ageing and deterioration which are very relevant to inspection planning for structures, as shown in Table 5.1.

As mentioned above, the current inspection tools are GVI, CVI and FMD and these methods cannot detect smaller cracks. Thus, it cannot be known whether widespread fatigue cracking in older structures which has not reached the through-thickness stage

Table 5.1 HSE guidance on management of ageing and thorough review for ageing installations.

Indicator of ageing	Examples relevant to offshore installations
External indicators of corrosion or deterioration	Paint blistering, rust streaks, evidence of corrosion at screwed joist or bolts, softening of passive fire protection (PFP). Surface corrosion of blast walls may indicate that their structural response has been adversely affected
External indications of incomplete reinstatement	Loose covers, ill-fitting enclosures, loose bolts, missing equipment, incomplete systems
Lack of commonality/ incompatibility	Replacement equipment of a later design or from an alternative supplier. Interface problems between modern and older control systems
Deterioration in structural performance	Initiation and propagation of fatigue cracks in structural members
Deterioration of uninspectable SCEs	For example foundations, ring stiffened and single sided joints
Increasing backlog of maintenance actions	An increase in the number of repairs that remain unresolved can be an indicator that ageing is taking place. As the maintenance backlog grows it can become increasingly difficult to get maintenance back on track
Inspection results	Inspection results can indicate the actual equipment condition and any damage. Trends can be determined from repeat inspection data. Water deluge performance parameters may be tracked through trend analysis
Experience of ageing of similar components	Unless active measures have been used to prevent ageing of similar components, it will be likely that the same problems can occur at other locations PFP is known to delaminate with age and structural movement, so the identification on one such problem is likely to indicate wider occurrence
Previous repairs	May indicate that ageing problems are already occurring, and since repairs have been needed during the life of the structure, the necessity of the repair will indicate the potential for further problems

Source: HSE 2009.

has occurred. This possibility needs to be considered as it could lead to significant overall damage occurring. Furthermore, a critical factor with techniques such as GVI, CVI and FMD is the frequency of inspection as it is important that through-thickness cracks cannot grow to cause local failure between inspections. Additional maintenance costs need also to be considered. An assessment of the redundancy of a structure should be performed to establish that failure of a small number of components, due to widespread damage, will not lead to structural collapse.

Some structural components are difficult or impossible to inspect but may be vulnerable to ageing processes and this presents a particular problem as their current condition is difficult to determine in terms of life extension. Such components could be in very deep water or with difficult access for an inspection tool. Examples are piled foundations where the circumferential butt welds are very difficult to inspect and ring-stiffened joints where the internal stiffeners are also difficult to inspect using conventional equipment. This inspection difficulty needs to be recognised at the design stage and current design practice is to provide such components with design fatigue factors ranging from 2 for non-critical components to as high as 10 for a critical component to reduce the likelihood of failure during their life. However, this was not properly recognised in early designs where a factor of two on life was the maximum used. This aspect needs to be taken into account in the fatigue assessment of an older structure. NORSOK N-006 gives guidance on the inspection of adjacent members and joints in order to achieve some information about an uninspectable detail.

5.2.6 Inspection of Floating Structures

Floating structures include a number of structural and marine systems that are different from fixed structures. The SCEs are the systems which control bilge and ballast, the watertight integrity, the weight control, stability monitoring and station keeping. The watertight integrity includes watertight doors and hatches, seals, valves, pumps, dampers, etc. Failure in any of these components can lead to loss of watertight integrity and possible loss of stability and buoyancy. Hence, this introduces a need for very different types of inspection and a very different way of thinking about the integrity management in general. In addition, the structural elements of a floating structure are quite different in geometry and form (as they are primarily shell structures, whereas fixed structures are often made of beams and tubular joints).

Many floating offshore structures in oil and gas production have a Class certificate which is provided by one of the leading classification societies, e.g. DNVGL and Lloyds Register. The inspection of floating structures has been a key part of the class system and regulation, which has been developed over many years of experience by owners, classification societies, etc. This is done on a calendar based system of inspection and surveillance and typically requires a dockside inspection every fifth year. Dockside inspection provides the opportunity for improved access and repair facilities.

Offshore floating structure used in oil and gas production typically stay on location for longer periods and the opportunity for regular dockside inspection is hence not available. This places a more onerous requirement for inspection of these permanently placed offshore floating structures, of which there is less experience. Several standards have recently been developed to guide the operators of floating facilities in their integrity management including inspection, such as ISO 19904-1 section 18

(ISO 2006) and NORSOK N-005 appendices F, G, H and I (Standard Norge 2017), which both follow the inspection procedure similar to that indicated in Section 5.2.2. In terms of ageing, this lack of experience could be significant and could require a more stringent inspection approach as there is an increased likelihood of deterioration being accumulated and widespread deterioration becoming more likely. In this regard, it should be mentioned that the draft API RP 2FSIM (API 2017) includes a separate annex specific to life extension which gives guidance on the life extension assessment process. However, it does not provide specific guidance on how to maintain these structures in the life extension phase after the assessment stage. In addition, API RP 2MIM (API 2018) is also highly relevant for the inspection of mooring lines.

In addition, API RP 2I (API 2015) and NORSOK N-005 (Standard Norge 2017) recommend inspection intervals for chain mooring systems for floating structures. In API RP 2I, the maximum interval between major inspections is linked to the age of the chain in years. For relatively new chains (i.e. 0–3 years), the recommended interval is 3 years; for slightly older chains (4–10 years), it is 2 years and for chains older than 10 years the interval is reduced to only 8 months. This short interval is very demanding and costly and hence chains are normally replaced before they reach the 10-year criterion. Inspection of mooring systems can be undertaken visually by an ROV but this has significant limitations. A more detailed inspection, e.g. by MPI, requires removal of the mooring and dockside inspection with significant cost and operational implications.

A document prepared by Oil & Gas UK (O&GUK 2014) entitled 'Guidance on the Management of Ageing and Life Extension for UKCS Floating Production Installations' provides detailed information relevant for inspection of ageing floating installations. Inspection features as a key control measure in managing the ageing mechanisms such as fatigue and corrosion, although little detail is given on the actual recommended inspection tools and technique. The main topics in the Oil & Gas UK document (O&GUK 2014) are the integrity management and inspection of:

- Hull (structural and watertight integrity).
- Marine system, including ballast system, control system, cargo system, inert gas system and marine utilities (pumps, generators, etc.).
- Station keeping systems (mooring and DP).

A comprehensive joint industry study on ROV inspection of mooring chains is reported in HSE (2017).

NORSOK N-005 builds upon this Oil & Gas UK report (O&GUK 2014) and breaks down the non-structural systems into a number of components and identifies whether inspection based on class rules or based on generally accepted maintenance standards such as NORSOK Z-008 (Standard Norge 2011) is appropriate.

5.2.7 Inspection of Topside Structures

Topside structures contain several safety critical components and their integrity needs to be managed to an appropriate level in the operational phase (including the life extension phase), reflecting that they suffer from degradation such as corrosion and fatigue.

The key components of a topside structure that needs to be managed are:

- The main load-bearing members of the deck structure (main frame, deck beam, primary structure of integrated decks, etc.).

- Flare booms.
- Helidecks.
- Derricks.
- Cranes and crane pedestals.
- Bridges.
- Support structures for risers, conductors, and caissons.[2]
- Load-bearing structures for modules.
- Supports for safety critical items, such as temporary refuge, living quarters, process equipment and piping, etc.

For all of these items, corrosion is usually the dominant degradation mechanism but fatigue must be considered also. Table 3.4 (in Chapter 3) shows the typical degradation mechanisms relevant for topside structures. The implication for inspection is that corrosion and fatigue need to be detected and this is increasingly important for ageing structures. On topsides, this can be done largely by using regular maintenance methods by the regular workforce. Depending on the access required this is relatively straightforward compared with subsea inspection.

Typically, problems occur when access is difficult, such as for structural elements under coatings, flare booms and derricks. Inspection under coatings is a particular problem as removal of the coating requires time-consuming reinstatement of the coating. Often, the reinstatement is local, creating potential problems with bonding to the original material.

In addition, coatings provide a key protection against corrosion and fire damage to the critical structural elements and hence need special attention in the inspection programme of topside structures. This is especially the case for passive fire protection (PFP) coatings.

PFP coatings are used on critical areas which could be affected by a jet fire. There are several different types which include cementitious and epoxy intumescent based. Maintenance of these coatings is very important because of the potential consequences of a fire on critical structural components. These coatings are inspected regularly as part of a maintenance programme. Typical damage includes water ingress, disbondment, surface cracking, spalling and erosion. This type of damage is more likely to be present in an ageing structure. An example of damaged PFP is shown in Figure 5.4.

The UK's HSE issued a special information sheet on inspection, evaluation and repair of damaged PFP (HSE 2007). This includes acceptance criteria for damaged PFP coatings which includes three severity levels. The most serious level (Level 1) is where gross failure of the PFP coating is found and where immediate remedial action is required involving removal and reinstatement of significant amounts of material. With Level 2 severity, some protection of the substrate remains but may reduce the fire resistance performance during a fire threat to a level that is unacceptable or is present in an area of high structural importance or will lead to significant further deterioration of the material. Remedial action will involve a repair requiring reasonable levels of reinstatement within an acceptable timescale.

Topside structure inspection is less well covered by standards compared with substructures. However, ISO 19901-3 (2010) was specifically prepared for topside structures, which includes a section giving a requirement for inspection of topsides. Also,

---

2 This book does not include detailed integrity management of risers, conductors and caissons.

Figure 5.4 Example of eroded PFP with retention mesh exposed but intact (HSE 2007).

NORSOK N-005 (Standard Norge 2017) gives specific requirements for topside structures.

In general, ISO 19901-3 (ISO 2010) refers to the requirements for in-service inspection and SIM given in ISO 19902 (ISO 2007). However, the former highlights corrosion as the most important deterioration process and the standard lists special topside specific areas requiring inspection such as access routes, floors, gratings and supports for safety critical equipment. In addition, the in-service inspection of main deck girders, bridges, flare booms, cranes, helidecks and other components where fatigue may be an issue follows from ISO 19902 (ISO 2007).

The standard indicates a default inspection scope from baseline to unscheduled, identical to the ISO 19902 standard, as described in Section 5.2.5. However, the standard does not describe a Level 4 inspection for topside structures, i.e. it implies that no detailed inspection is required. However, Level 3 includes a detailed non-destructive examination of all safety-critical structural components. The periodic inspection described in ISO 19901-3 highlights PFP and coatings as very important for topsides, as mentioned above.

HSE (2000) reviewed standards for both fabrication and in-service inspection of topsides. It was concluded that there should be a more systematic approach of the assessment of the structural system interaction with the process plant and pipework than the approaches generally identified within current codes and standards at that time (it should be mentioned that the authors of this HSE report had access only to an early draft of the ISO 19901-3). This interaction is important because should the support structure fail there is a potential of hydrocarbon leakage that could lead to fire and explosion. The report also concludes that one possible way forward would be to develop a quantitative risk based assessment of inspection class for topside components. This would permit a rational and consistent approach to be adopted. Since this report was published, there have been developments in risk based inspection of topside structures and many engineering companies are able to deliver such inspection plans to operators and duty holders. However, the implementation in practice is still dependent on the individual operator and a more general introduction of such an approach will most likely require inclusion of this into standards such as ISO 19901-3.

5.2.8 Structural Monitoring

Structural monitoring (also called on-line instrumentation, on-line monitoring or structural heath monitoring) is instrumented observation of a structure using sensors that continuously or periodically measure structural behaviour to identify changes to material or geometric properties and boundary conditions of the structure. Structural monitoring includes use of analysis (statistical pattern recognition) of the data to identify changes that indicate damage or other changes to the structure. The most used structural monitoring methods presently in use are:

- Natural frequency response monitoring.
- Leak detection.
- Air gap monitoring.
- Global positioning system monitoring.
- Fatigue gauge.
- Mooring chain tension monitoring.
- Acoustic emission monitoring.
- Acoustic fingerprinting.
- Strain monitoring (e.g. strain gauges or optical fibres).

There is good evidence that the existence of the damage on the structure, and in some cases the location of the damage, can be detected by structural monitoring. The type of damage and its severity may also be defined by the most advanced monitoring and analysis methods (May et al. 2008).

Structural monitoring can complement existing inspection techniques to provide greater confidence in structural integrity or to reduce inspection cost. It can assist in meeting the requirement to manage lifetime structural integrity, as well as providing information to demonstrate the case for life extension. Structural monitoring can, in combination with other methods, be used to demonstrate continued safe operation of a structure beyond its original design life. Other typical applications for structural monitoring include:

- Monitoring of a known local defect or high-risk part of a structure.
- Reduction of the NDT/NDE inspection activity (under certain circumstances).
- Demonstration of compliance with regulatory requirements.

On-line instrumentation has been used to benefit the development of offshore oil and gas structures. Projects such as Shell's Tern and BP's Magnus have provided a much better understanding of the loading on these platforms, enabling more efficient designs to be developed for future structures (HSE 2009 RR685). Monitoring of damage on offshore structures, such as using acoustic emission on the Ninian Southern platform (Mitchell and Rodgers 1992), gave confidence that operations could continue whilst repairs were being planned and undertaken. A useful source of information is the Simonet structural integrity monitoring network and web site (www.simonet.org.uk).

More information on structural monitoring is given in a study commissioned by the HSE (HSE 2009 RR685). The topics of this study included:

- Identification and assessment of current structural monitoring technologies applicable for use offshore.
- Review of relevant codes and standards and other background literature.
- A summary of the current status of structural integrity monitoring in the offshore oil and gas industry.

For certain purposes, structural monitoring can be used as a cost-effective alternative to conventional inspection methods, particularly for monitoring areas with limited accessibility. It is also an important tool to verify novel design solutions (NORSOK N-005 1997).

A number of offshore codes and standards make reference to structural condition monitoring. These include ISO 16587 (ISO 2004), ISO 19902 (2007), API RP2 SIM (API 2014) and NORSOK N-005 (Standard Norge 2017). In most cases, the reference indicates that structural condition monitoring can be used as an addition to normal inspection. These standards include structural condition monitoring systems in the following situations:

- ISO 19902 indicates that where air gap measurement devices are correctly set up, calibrated and maintained, continuous records of wave heights and tide can provide very useful information on environmental conditions. Where this can be combined with directionality data and ideally some method of estimating actions (e.g. strain gauges), the data can be used in analyses and assessment of defects and of remaining life, possibly reducing conservatisms. Further, ISO 19902 indicates that satellite surveying techniques can often be used to determine the position of the structure and hence e.g. the air gap.
- API RP2 SIM indicates that the monitoring of fatigue sensitive joints and/or reported crack-like indications may be an acceptable alternative to analytical verification.
- ISO 16587:2004 describes the performance parameters for assessing the condition of structures, including types of measurement, factors for setting acceptable performance limits, data acquisition parameters for constructing uniform databases, and internationally accepted measurement guidance (e.g. terminology, transducer calibration, transducer mounting and approved transfer function techniques).

Table 5.2 indicates the range of parameters that can be monitored with current techniques. More details on each of these techniques are provided in Appendix B.

It is clear from the above that the offshore industry has implemented several techniques during the last few decades to monitor (continuously) the condition of offshore structures. In general, these have been applied on a one-off basis as these techniques are very varied in terms of maturity and applicability. For these reasons, continuous structural monitoring has not been widely taken up by the industry, which has rather favoured periodic inspection schemes instead.

As structures are increasingly called upon to operate beyond their original design life, there is a gap between the knowledge about the structure and the actual structural integrity. This 'knowledge gap' is caused because the likelihood of individual or combinations of fatigue failures may increase significantly beyond design life. Therefore, the possibility of the structure existing in an unknown unsafe condition also increases. One way of closing the knowledge gap is by the implementation of a system that continuously monitors for structural degradation.

Table 5.2 Capability of monitoring techniques.

Structural monitoring technique	Monitoring technique examples	Monitoring capability
Air gap monitoring	GPS, laser	Loss of air gap
Global positioning system	GPS	Loss of station keeping
		Loss of air gap
Acoustic emission	Acoustic stress wave monitoring devices	Fatigue crack initiation
		Fatigue crack growth
		Corrosion
Continuous flooded member detection and leak detection	Electrical detector cells activated by water or buoyancy based	Member leak detection
		Breach of water tight integrity
		Through-thickness cracking
		Through-thickness corrosion
Natural frequency response monitoring	Accelerometers	Member severance
		Significant damage to member and joints
		Damping
Fatigue gauge	Strain gauges	Fatigue cracking
Acoustic fingerprinting		Through thickness cracking (fully severed members)
Strain monitoring		Local stress and loading regime

Continuous structural monitoring provides an opportunity to monitor the degradation of ageing structures where there is uncertainty about the condition of the structure from using periodic inspection. Further, as mentioned in Chapter 4, it is of particular importance to have early warning signals from ageing structures. One good option for providing such early signals would be undoubtedly from continuous condition monitoring. This requires that the likely and the critical failure modes of the structure are well understood and that the structural condition monitoring system is designed to be capable of identifying such failures.

5.3 Evaluation of Inspection Findings

The inspection and other surveillance of the structure will produce new data about the current condition and configuration, trends in any degradation and the loading on the structure. When such new data become available, it is necessary to undertake an evaluation of the structure (see also the SIM process in Section 2.4) to decide whether it is sufficiently safe and fit for purpose up to the time of the next planned inspection, or if:

- Immediate actions are needed (if the data indicate that the structure is in immediate danger of failing).
- There are trends in any degradation mode.
- Mitigating measures are needed, such as repair, strengthening or weld improvement.
- Further analysis (assessments) are needed.
- Further inspections are needed.
- The existing surveillance programmes are adequate and properly executed.

The evaluations will often be performed by checking against predefined acceptance criteria, for example crack size, corrosion extent, acceptable loading on decks, etc. Methods for determining such acceptance criteria for cracks are indicated in Section 4.6.

Some simplified calculations may be performed as a part of the evaluation but if an analysis is needed this is normally done as a part of an assessment. If such analysis is needed it is important that the results of the evaluation (accurate information about the anomalies) is communicated to engineers undertaking the assessment. Evaluation also needs to include preparation of the necessary documentation for the execution of corrective actions and mitigations if needed.

An anomaly found during inspection and surveillance may require changes in the future inspection and surveillance programme to monitor and record the status of the affected area and any remedial or mitigating measures if relevant. Hence, the evaluation engineers should ensure that such information is included in the long-term inspection and surveillance plan.

Evaluation requires consideration of numerous factors affecting the structural performance and corrosion protection for the various structural components. ISO 19902 (ISO 2007) indicates the following structural performance factors that need to be considered:

- Age of the structure, its location, current condition, original design situations and criteria and comparison with current design criteria.
- Analysis results and the assumptions for the original design and subsequent assessments.
- Structural reserve strength, structural redundancy and fatigue sensitivity.
- Degree of conservatism or uncertainty in specified environmental conditions.
- Previous in-service inspection results and learnings from performance of other structures.
- Modifications, additions, and repairs or strengthening and presence of any debris.
- Occurrence of any accidental and severe environmental events.
- Criticality of the platform to other operations.

In terms of corrosion control, ISO 19902 (ISO 2007) indicates the following aspects to be considered in the evaluation:

- The assumptions and criteria used in design.
- The details of the system (impressed current or sacrificial anodes) and its past performance.
- CP readings from monitoring, compared with design criteria.
- State of the anodes from visual inspection (if a sacrificial system is used).

If the evaluation results are unacceptable, in terms of not meeting the criteria described above, then further analysis (assessment) may be required. Alternatively, mitigation measures which reduce the likelihood of structural failure need to be implemented as described in the next section.

5.4 Mitigation of Damaged Structures

5.4.1 Introduction

Mitigation and structural repairs to offshore installations may be required when damage is found during inspection. Mitigation and repair are associated with the reduction of potential failure for structures that are damaged or degraded and can involve repair, strengthening or, in some cases, monitoring until further actions are deemed necessary.

Whether mitigation is required or not is determined through an evaluation process, see Section 5.3. If the evaluation (and possible additional assessment process) indicates that mitigation is required, a number of different approaches and methods are available, depending on the type of damage and degradation. The choice of remedial measures and their extent will depend on the source of risk to structural integrity and its magnitude.

Mitigation techniques such as strengthening, modification and repair generally include (Dier 2004):

- Welding (dry or wet welding).
- Clamp technology (mechanical clamps, grout filled clamp/sleeves and neoprene-lined clamps).
- Grout filling (of members or joints).
- Weld improvement (to grinding, profile grinding or hammer peening).
- Other (remedial grinding of weld, composites, member removal and bolting).

Weld improvements methods may be categorised as (Haagensen and Maddox 2006):

- Machining methods (burr grinding, disc grinding and water jet gouging).
- Re-melting methods (TIG dressing, plasma dressing and laser dressing).
- Residual stress modification methods:
  - Mechanical peening methods (hammer peening, needle peening, shot peening, ultrasonic peening).
  - Mechanical overload methods (initial overloading, local compression).
  - Thermal methods (thermal stress relief, spot heating, Gunnert's method).

These weld improvement methods are primarily used to extend the fatigue life of welds and the best results (Haagensen and Maddox 2006) are obtained by a combination of a machining method to modify the weld (e.g. grinding) and a residual stress modification method (e.g. peening).

Structural repairs and strengthening to fixed steel jacket structures have generally been either to repair a cracked welded component or provide an alternative load path (e.g. mechanical or grouted clamps). There is a considerable amount of planning required prior to an underwater repair, sometimes requiring assembly of the components on land to minimise the time required underwater. In this respect, the ability to have access to joints or members for repair is a key factor in designing how a repair is implemented; this accessibility is determined at the design stage. In some respects, the accessibility is also relevant to inspectability (see previous section) and more likely to be considered during design.

Fixed steel jacket structures contain a range of different components, some of which are more easily repairable than others – see Table 5.3.

It can be seen that the most difficult components to repair are piles and grouted connections. A further important factor is that these components are also very difficult to inspect. It is important therefore that sufficient capacity is provided at the design stage such that any expected reduction in strength as a result of ageing effects can be allowed for, recognising the difficulty of repair.

Table 5.3 Repair capabilities for components on a fixed steel jacket structure.

Structural component	Typical damage	Ease of repairability
Tubular member	Denting, bowing, corrosion	Grout filling of members. Members can be replaced or repaired locally, access is a factor
Tubular joint (welded)	Fatigue cracking	Cracks can be repaired (hyperbaric welding) requiring complex habitat or alternative load path provided by, for example, grouted clamp type repair. Grinding also of the cracked area for small cracks to improve fatigue life
Foundation piles	Scour, loss of pile capacity due to pile corrosion, soil degradation	Scour can be mitigated by use of e.g. rock dumping. Loss of pile capacity very difficult to repair
Grouted connections	Loss of connectivity with piles	Very difficult to repair, provision of alternative load path main option
Sacrificial anodes (for CP)	Usage during life, mechanical damage	Anodes can be replaced or alternative CP system installed
Riser supports	Corrosion, fatigue	Weld repair, local grouted repairs

Possible mitigation measures for both corrosion and fatigue damaged structures are now considered.

5.4.2 Mitigation of Corrosion Damage

As noted in Section 3.4, management of corrosion in offshore installations includes the following zones:

- Underwater: provision of a corrosion protection system such as sacrificial anodes, impressed current or anti-fouling (ship-shaped structures).
- Splash zone: use of coatings and corrosion allowance (additional steel thickness).
- Atmospheric zone (including topsides): mainly by protection using coatings.

Even with these protective systems in place, most offshore installations experience some degree of corrosion. Typical examples are due to damaged coatings or poor control of the protection voltage.

Remedial measures include:

- Improvements to the corrosion protection system.
- Repair of damaged coatings.
- Patch repairs (removal of corroded section and replacing with patch).
- Replacement with new steel member or structural part.

5.4.3 Mitigation of the Corrosion Protection System

Corrosion protection system systems are designed at the beginning of the life of the installation, with an expected life commensurate with the design life. Life extension therefore may require review of these systems and consideration of any mitigation required.

For underwater systems, sacrificial anodes are designed with usually a 25-year life based on the expected current demand of the system. As part of the inspection plan, the anode condition is checked and, depending on the remaining material, replacement may be required. Figure 5.5 shows a series of drawings of sacrificial anodes with increasing loss of material. At some stage (e.g. the severely pitted example) this results in a less negative protection potential and therefore less efficient CP. At this stage, anode replacement is required to maintain efficient protection.

Individual anodes can be replaced but this can involve a considerable amount of working underwater, often requiring divers. An alternative is to provide a pod containing several anodes to upgrade the protection in a local area.

These pods are deposited on the seabed and connected using a clamp tieback system to any tubular member or flange, allowing for less installation time with little or no diver intervention. However, this system can result in less efficient spread of protection than by using conventional individual anodes.

An alternative approach is to use a bracelet containing several anodes which can be clamped around a member on a platform.

Removal of an installation from the sea as part of decommissioning allows the state of anodes to be examined in more depth. One example was the recovery of the West Sole WE platform after 11 years in the sea. A detailed examination of the several components was undertaken including the anodes, as shown in Figure 5.6. The weight loss of the anodes was found to be 40%, less than the expected value. Protection potentials were measured before recovery and were in the range 0.87–0.93 V compared with Ag/AgCl, which is more positive than the level set at the design stage. The splash zone coating was a coal tar epoxy and it was found that the protection had broken down from wave and spray action. It was noted that this could have been minimised by more painting during maintenance. Isolated pitting and metal wastage was seen on members in the

Anode slightly pitted and with rounded edges

Figure 5.5 Examples of anode condition from ageing.

Anode moderately pitted, approximately 50% missing

Anode severely pitted where frame is visible

Figure 5.6 Aluminium anode recovered from West Sole WE platform showing less wastage than expected, with an average weight loss of 40%. Source: J.V. Sharp.

splash zone, up to a depth of 3 mm in places, which could have become more serious if the platform had been in the sea for a longer period.

Coatings are applied for corrosion control because they act as barriers that isolate the metal from the corrosive environment. Coatings can be applied in the construction yard or offshore although the latter is more expensive and often less efficient. Typical coatings used offshore include coal tar epoxy (as noted above for the West Sole WE platform) and epoxy resins. In addition, PFP coatings are used on areas of a platform potentially subject to jet fires. Coatings are inspected as part of a maintenance programme and may require repair or replacement depending on the level of damage. Several scales exist (ISO 2011) for measuring the degree of either blistering, rusting, flaking or cracking which may occur with coatings. Pinpoint rusting is one of the more common types of coating failure, often as a result of ageing. Assessing the level of damage to a substrate can be difficult without removing the coating, which can be expensive and require substantial remedial work.

Coatings and anodes are also a feature of corrosion protection for ballast tanks in floating structures. Breakdown of these coatings can lead to localised corrosion and loss of water integrity. Proper re-instatement of coatings in such areas is an important mitigating measure.

5.4.4 Mitigation of Fatigue and Other Damage

Structural repairs are often required when fatigue damage is found during inspection and analyses show that this damage, unless repaired, will reduce the safety level of the structure. There are a number of different structural repair methods which include two main types:

- Local modifications to the damaged weld.
- Providing an alternative load path or strengthening the damaged member.

Examples of local modification are repair welding, drilling holes at both ends of a crack and weld grinding. Repair welding would involve the removal of cracks by grinding and the resulting excavation filled by manual metal arc welding or burr grinding to achieve a smooth finish on the repair weld face. Hole drilling and cold expansion involves removing the fatigue crack tips by drilling through-thickness holes, which are then cold expanded to include compressive residual stresses in the hole circumference. Grinding alone can be used to remove cracks penetrating partly through the wall thickness by burr grinding and the resulting excavations left unrepaired. All these methods have been studied with respect to the performance of the repaired structures, which is further discussed in Section 5.5.2. In general, it has been found that the fatigue strength of as-welded repair welds is marginally lower than that of unrepaired welds.

Examples of providing an alternative load path to the damaged member are clamps (grouted and mechanical strengthening), grout filling of damaged tubular members and local strengthening by the adding of brackets or improving the geometry of brackets at crack prone details. A typical example of this would be adding improved brackets at side-longitudinal connections with transverse frames or bulkheads.

For each type of repair, the factors that need to be addressed include:

- Its application and operational requirements, particularly for underwater repairs.
- Ranges of application.
- Permissible working loads.
- Static strength.
- Fatigue.

Techniques for weld repair include hyperbaric welding (welding underwater using a habitat to maintain dry conditions) and wet welding (which generally has problems in providing the strength of a weld made in the air). The former is an expensive process involving building a habitat underwater and using divers in the chamber to effect the weld repair but is capable of achieving good properties. The use of divers also requires expensive support vessels.

Typical ways of providing an alternative load path is by grout-filling the member or by using structural components such as mechanical connections and clamps. There are several examples where these have been successfully achieved offshore (Dier 2004). However, the cost is high because of the required underwater working and support vessels. A typical structural repair is shown in Figure 5.7.

Several research programmes (Department of Energy 1988a,b) have been expedited to provide design information for such repairs, including both fatigue and static strength properties. These have included testing of repaired and grout-filled components in the laboratory. One of the critical factors in repairs is being able to obtain a close connection with existing steelwork such that stress transfer can be effective. Ways to achieve this have included using grout to fill the interface between the original steelwork and the clamp or by using an elastic membrane (e.g. neoprene). The stress transfer can be made

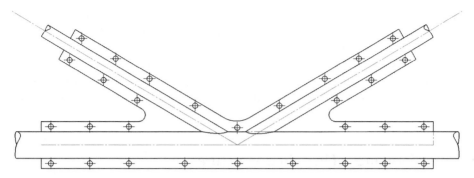

Figure 5.7 Schematic diagram of a repair of a cracked tubular K joint using bolted connections.

more effective by stressing the clamp directly onto the tubular section by using long studbolts (Dier 2004). Structural repairs and clamps have performed satisfactorily over many years but require regular inspection particularly if bolts have been used to connect the two halves of a repair. These repairs and clamps have a disadvantage in increasing the wave loading, a factor which needs to be taken into account during design of the repair and the assessment of the global structure.

Grout filling of damaged members or joints can increase their static strength, particularly if the members have suffered impact damage such as denting. In terms of fatigue, the introduction of grout considerably reduced the stress generated under axial and out-of-plane bending conditions, thereby reducing the stress concentration factors (SCFs) around the brace/chord weld intersection. It was also shown that the failure modes of grout stiffened nodes of the geometry investigated in the test programme had not been altered from that of conventional nodes.

Figure 5.8 also shows a bolted clamp used to strengthen a connection on the Viking AD platform. This clamp was recovered from the platform following decommissioning which allowed tests to be carried out on the components such as the bolts and grout.

Figure 5.8 Bolted clamp used on the Viking AD platform, following decommissioning. Source: J.V. Sharp.

In semi-submersible and ship-shaped structures, repair of fatigue damage is often done by grinding and weld repair, which is much easier if performed in dry conditions (e.g. dry dock). The finding of a fatigue crack in a semi-submersible or ship-shaped structure often involves bad detailing in the design or fabrication of, for example, stiffeners. Remedial measures in these cases often include redesign of the detail to reduce the stress concentration factor or grinding the existing stiffener in order to reduce the effect of the hot-spot stress.

5.5   Performance of Repaired Structures

5.5.1   Introduction

Repaired structures improve overall performance, especially for strength and corrosion, but typically continue to have some fatigue issues. The safety of a repaired structure depends on the quality of the design and fabrication, as well as the workmanship used in making the repairs. Therefore, these repairs require follow-up inspection after the repair has been in operation for a certain period.

5.5.2   Fatigue Performance of Repaired Tubular Joints

A study (Department of Energy 1988a,b) entailing a series of fatigue tests on welded tubular T joints in which fatigue cracks were repaired by a number of alternative methods aimed to establish a ranking for repair methods in terms of residual fatigue performance and to establish whether it is necessary to repair the entire joint or only the cracked region. The joints were tested in out-of-plane bending. The repair methods investigated were:

- Repair welding (cracks were removed by grinding and the resulting excavation filled by manual metal arc welding).
- Repair welding and burr grinding (repair welds, made as described above, were fully burr-ground to remove the repair toe weld and to achieve a smooth finish on the repair weld face).
- Hole drilling and cold expansion (the fatigue crack tips were removed by drilling through-thickness holes, which were then cold expanded to include compressive residual stresses in the hole circumference).
- Grinding alone (part-wall cracks were removed by burr grinding and the resulting excavations left unrepaired).

Repair welding was carried out in air at atmospheric pressure under shop conditions rather than trying to reproduce conditions likely to be met in repairing an offshore structure. In a subsidiary investigation, fatigue crack growth rates were measured in weld metal deposited under hyperbaric conditions to establish whether hyperbaric repair welds would give similar results to those for one-atmosphere repairs (Department of Energy 1989). It was found that:

- Expressed in terms of the hot spot stress range, the fatigue strength of as-welded repair welds was marginally lower than that of unrepaired welds. However, this was compensated for by the fact that in making the repair the overall weld length was increased which led to a reduction in the stress range of the repair weld toe. As a result, if the applied loading is of the same magnitude after repair as before, the fatigue endurance for through-thickness cracking after repair is similar to or marginally greater than that before repair.

- Burr grinding the repair weld could considerably improve the fatigue strength, although a smooth surface finish needed to be achieved to avoid premature crack initiation on the weld face. Where the applied loading continued unchanged after the repair, the fatigue life to through-thickness cracking was in excess of twice that before repair.
- Through-thickness and part-wall weld repairs behaved similarly. Under the mode of loading investigated, a relatively large number of penetrating defects could be tolerated at the root in through-thickness repairs.
- Cold expanded holes at the crack tips were not effective as a means of delaying crack propagation.
- Removal of part-wall flaws by grinding was an effective repair method giving endurances after repair up to four times greater than the mean for unrepaired joints. Data were presented which allow the likely residual life to be estimated for a given excavation depth.
- It was necessary to burr grind the unrepaired weld toes on both the chord and brace side to avoid premature failure.
- The crack growth rates in hyperbaric repair welds would be expected to be similar to those in the one-atmosphere repairs studied in this report.

The Joint Industry Repairs Research Project (Department of Energy 1988a,b) was funded by the UK Department of Energy and nine oil companies. The research carried out investigated the static strength and fatigue performance of grouted and mechanical connections and clamps of the types used to strengthen or repair underwater steel members. A further study (Department of Energy 1989) presented details and results of a test programme carried out on tubular members to determine the effectiveness of grout filling as a means of repairing damage. The tests involved indenting a number of grout-filled specimen lengths under controlled conditions. After denting the tubular were tested and then after failure cut open for the condition of the grout to be examined. Comparisons were made with the unfilled damaged and undamaged parent tubulars. A comparison was also made with the experimental work and theoretical analysis undertaken separately.

It was shown that for the number and dimensions tested the presence of grout enhanced the strength of the member when compared with an identical, but unfilled, damaged tubular. The results show that this increase in strength is dependent upon grout strength, dent size and L/r (length to radius of gyration ratio) and D/t (diameter to thickness ratio). At higher grout strengths, increases in ultimate strength of between 45% and 125% were achieved. The size of dents (d) influences the strength with an increase of approximately 70% for small dents ($d/D = 0.04$) and an increase of 55% for larger dents ($d/D = 0.16$). The L/r ratio has little effect at low values of the D/t ratio but a reduction is observed at higher D/t ratios.

The strength increased with D/t ratio for lower values of the slenderness ratio (L/r) but there was little change at higher ratios. Except for the most severe case of damage ($d/D = 0.16$) tested, the presence of high strength grout increased the strength of a damaged member to beyond that of the same unfilled undamaged member. The increase in strength, however, resulted in post-collapse losses and rapid unloading. Scale effects would appear to be present with smaller loads being associated with smaller scale tests, although these effects do not change the main conclusion of enhanced strength from grout filling. From the limited test results, it would appear that fatigue is not a significant problem at acceptable grout strengths. Since the ultimate load was found to be influenced by grout strength in the static case, fatigue may be an important consideration at lower grout strengths.

The report (Department of Energy 1992) describes static stress analysis and fatigue tests on repaired and fully internally grouted tubular welded T joints. Two fatigue damaged 914 mm diameter tubular T joints originating from the UKOSRP programme were repaired and internally grouted. Extensive stress analysis was carried out on each specimen to determine the influence of full internal grout on the stress and strain concentration factors generated under three-point loading, axial loading through the brace member, in-plane bending and out-of-plane bending. Each specimen was then subjected to axial fatigue loading through the brace members to evaluate the effect on fatigue performance and failure modes due to the introduction of this type of stiffening. Throughout each test, crack initiation sites and crack propagation data were recorded along with joint flexibility data (Department of Energy 1989).

The results presented in the form of stress concentration factors and stress endurance curves indicated that the technique could be applied to existing nodes to extend their fatigue lives by reducing the hot spot stresses around the chord/brace intersection due to the loads which the nodes are subject to in service. In addition to the possible extension in fatigue lives, crack propagation and local joint flexibility indicated that failure modes had not changed from those exhibited by conventional nodes. More detailed results were:

- Introduction of grout considerably reduced the stress generated under axial and out-of-plane bending conditions thereby reducing the SCFs around the brace/chord weld intersection weld.
- The fatigue lives of two grout stiffened nodes were lower than predicted by the mean T curve but are within the scatter band of that displayed by conventional nodes.
- Failure modes of grout-stiffened nodes of the geometry investigated in this programme have not been altered from that of conventional nodes.
- Fatigue lives of existing undamaged nodes could be extended by the introduction of grout provided the dominant load condition is axial.
- The existing *S–N* curve approach to structural analysis was still applicable for grout-stiffened nodes.
- Load shedding of a grouted node was similar to that of conventional nodes.

5.5.3 Fatigue Performance of Repaired Plated Structures

According to Ship Structure Committee (SSC) report 425 (2003), little guidance and standards are available on the fatigue performance of repaired plated structures in ship type structures. However, the TSCF (Tanker Structural Cooperative Forum) provides several sources on the condition evaluation and repair of typical cracks in ship-shaped structures (TSCF 2018):

- Condition Evaluation and Maintenance of Tanker Structures, 1992.
- Guidelines for the Inspection and Maintenance of Double Hull Tanker Structures, 1995.

- Guidance Manual for Tanker Structures, 1997.
- Guidance Manual for Maintenance of Tanker Structures, 2008.

The SSC report further indicates that repair techniques can be categorised into three major categories:

1. Surface crack repairs.
2. Repair of through-thickness cracks.
3. Modification of the connection or the global structure to reduce the cause of cracking.

References

API (2014). API RP-2SIM Recommended Practice for Structural Integrity Management of Fixed Offshore Structures, American Petroleum Institute.

API (2015). API-RP-2I In-service Inspection of Mooring Hardware for Floating Structures, 3e, American Petroleum Institute.

API (2017). API-RP-2FSIM Floating Systems Integrity Management – Draft, American Petroleum Institute.

API (2018). API RP-2MIM Mooring Integrity Management – Draft, American Petroleum Institute.

CSWIP (2018). Certification scheme for personnel compliance through competence. www.cswip.com (accessed 5 April 2018).

Dalane, J.I. (1993). System reliability in design and maintenance of fixed offshore structures. PhD thesis. Norwegian Institute of Technology, University of Trondheim, Norway.

Department of Energy (1988a). Offshore Technology Report – Fatigue Performance of Repaired Tubular Joints, OTH 89 307, HMSO.

Department of Energy (1988b). Grouted and Mechanical Strengthening and Repair of Tubular Steel Offshore Structures, OTH 88 283, HMSO.

Department of Energy (1989). Offshore Technology Report – Residual and Fatigue Strength of Grout Filled Damaged Tubular Members, OTH 89 314, HMSO.

Department of Energy (1992). Fatigue Life Enhancement of Tubular Joints by Grout Injection, OTH 92 368, HMSO.

Dier, A.F. (2004). Assessment of repair techniques for ageing or damaged structures. MMS Project #502, Funded by Mineral Management Service, US Department of the Interior, Washington, DC under contract number 1435-01-04-CT-35320.

DNVGL (2015). DNVGL-RP-C210 Probabilistic methods for planning of inspection for fatigue cracks in offshore structures. DNVGL, Høvik, Norway.

Dover, W.J. and Rudlin, J.R. (1996). Defect characterisation and classification for the ICON inspection reliability trials. *Proceedings of 1996 OMAE*, vol. II, pp. 503–508.

Haagensen, P.J. and Maddox, S.J. (2006). IIW recommendation on post weld improvement of steel and aluminium structures. IIW report XIII-1815-00, International Institute of Welding.

HSE (2000). HSE OTO 2000/027. Review of current inspection practices for topsides structural components, Health and Safety Executive (HSE), London, UK.

HSE (2007). Information Sheet Advice on acceptance criteria for damaged Passive Fire Protection (PFP) Coatings. Offshore Information Sheet No. 12/2007, Health and Safety Executive (HSE), London, UK.

HSE (2009). Information Sheet Guidance on management of ageing and thorough reviews of ageing installations. Offshore Information Sheet No. 4/2009, Health and Safety Executive (HSE), London, UK.

HSE (2017). Research report RR1091, Remote Operated Vehicle (ROV) inspection of long term mooring systems for floating offshore installations, Health and Safety Executive (HSE), London, UK.

ISO (1998). ISO 2394:1998, *General principles on reliability for structures*, International Standardisation Organisation.

ISO (2004). ISO 16587:2004, *Mechanical vibration and shock – performance parameters for condition monitoring of structures*, International Standardisation Organisation

ISO (2006). ISO 19904:2006, *Petroleum and natural gas industries – floating offshore structures*, International Standardisation Organisation.

ISO (2007). ISO 19902, *Petroleum and natural gas industries – fixed steel offshore structures*, International Standardisation Organisation.

ISO (2010). ISO 19901-3:2010, *Petroleum and natural gas industries – specific requirements for offshore structures – Part 3: Topsides structure*, International Standardisation Organisation.

ISO (2011). ISO 4628, *Evaluation of degradation of coatings*, International Standardisation Organisation.

ISO (2013). ISO 19900:2013, *Petroleum and natural gas industries – general requirements for offshore structures*, International Standardisation Organisation.

ISO (2017). ISO/DIS 19901-9:2017, *Structural integrity management*, International Standardisation Organisation.

May, P., Sanderson, D., Sharp, J.V., and Stacey, A. (2008). Structural integrity monitoring – review and appraisal of current technologies for offshore applications, Paper OMAE2008-57425. In: *Proceedings of the 27th International Conference on Offshore Mechanics and Arctic Engineering,* Estoril, Portugal (June 2008). New York: American Society of Mechanical Engineers.

Mitchell, J.S. and Rodgers, L.M. (1992). Monitoring structural integrity of North Sea production platforms by acoustic emission, Offshore Technology Conference, OTC-6957-MS.

O&GUK (2014). Guidance on the Management of Ageing and Life Extension for UKCS Floating Production Installations, Oil and Gas UK, London, UK.

SSC (2003). Fatigue strength and adequacy of weld repairs, Ship Structure Committee report no. 425.

Standard Norge (2011). NORSOK Z-008 Risk based maintenance and consequence classification – Rev. 3. Standard Norge, Lysaker, Norway.

Standard Norge (2015). NORSOK N-006 Assessment of structural integrity for existing offshore load-bearing structures. 1e. Standard Norge, Lysaker, Norway.

Standard Norge (2017). NORSOK N-005 In-service integrity management of structures and maritime systems. 2e. Standard Norge, Lysaker, Norway.

TSCF (2018). Reports on TSCF. www.tscforum.org (accessed 4 April 2018).

6

Summary and Further Thoughts

6.1 Ageing Structures and Life Extension

As shown in this book, ageing involves offshore structures being exposed to an environment causing progressive degradation and damage from loads from natural hazards and accidental impacts. The way these structures are used will often change over the years, which as a result will modify the loading and possibly the configuration of the structure. Such changes in condition, loading, configuration or knowledge about the structure can, unless addressed, undermine confidence in its integrity and acceptability for further service within evolving technological and regulatory regimes.

The continued need of older offshore structures for the production of oil and gas means that extension of their life beyond the original design specification is essential. Beyond the life extension stage, decommissioning and removal of structures could take many more years after production has ceased. Hence, the changes due to ageing outlined above need to be addressed, understood and taken into account to enable the structural integrity to be maintained and demonstrated during the extended life.

These changes have been addressed in the different sections of this book with the aim of demonstrating how structures and confidence in their integrity can be maintained as they age and get older. In addition, the book has indicated methods for inspecting, assessing and evaluating these structures properly for the safe management of these older structures in their life extension phase. The book has highlighted the importance of increased knowledge and history and ensuring that structures that are found unfit for further service are upgraded or decommissioned.

The book has recognised that ageing and life extension can introduce new issues and challenges that need to be addressed. These include:

- The likelihood of damage and degradation will increase with time and, hence inspection is to a large extent more significant for the continued safety of the structures, increasingly requiring a higher level of inspection capability than some current methods provide.
- Widespread degradation and damage (involving several occurrences at adjacent sites in the structure) are more likely for ageing structures, which highlights an increased need for inspection but also for structural redundancy and ductility.
- Uncertainties about the state of a structure which increase with time to an extent that reduces confidence in its integrity and has to be dealt with by the assessment methods described in this book.

Ageing and Life Extension of Offshore Structures: The Challenge of Managing Structural Integrity,
First Edition. Gerhard Ersdal, John V. Sharp, and Alexander Stacey.
© 2019 John Wiley & Sons Ltd. Published 2019 by John Wiley & Sons Ltd.

- The available information on the history of the structure may not be readily available as a result of organisational changes and poor record-keeping over the years.
- Engineering methods used to analyse the structures and the standards used may have changed over the years since the structure was installed.
- Obsolescence may become important for older installations and should be considered, especially for the marine systems involved with stability, ballasting, station keeping, and water tight integrity of floating structures.
- As a result of the issues raised above, a structure that is being life extended clearly needs to be inspected and hence the inspectability of the structure is a vital requirement; areas that are difficult to inspect provide a particular challenge.
- Repairability of the structure is a vital requirement and areas that are difficult to repair also provide a particular challenge.
- As expertise in life extension develops, with the increasing number of ageing structures, improved assessment methods will be developed and need to be included in standards.
- Economic considerations, including uncertainties about continued operations, may influence the management of life extension.

Increasing recognition of the importance of structural integrity management (SIM) has led to the emergence of standards on SIM in recent years (e.g. by American Petroleum Institute [API], International Standardisation Organisation [ISO] and NORSOK). These standards are a first step in establishing consistent and good industry practice. However, the development of guidance and industry practice on life extension to enable wider awareness of ageing issues is still required in the offshore industry. The ever-increasing number of ageing structures places increasing importance on this activity.

6.2 Further Work and Research Needs Related to Ageing Structures

Research and technological advances have historically played an important part in developing industry practice and standards for the design and operation including inspection and assessment, of offshore structures. As structures age there is a continuing need for further development of these standards and industry practices. The book has identified a number of areas associated with ageing structures that could benefit from further research.

In recent years, the understanding of key degradation mechanisms has developed considerably due to the availability of more research, better test data and experience from operations. However, there still remain a number of issues which require further work in relation to ageing offshore structures. These include:

- The treatment of components that are either very difficult or impossible to inspect in service, such as complex nodes, underdeck structures, piles and foundations. The integrity of these depends on assessment of design data, contemporary assessment procedures and advancing inspection and monitoring technology. For components which are more critical, e.g. pile connections for fixed structures, an assessment needs to be performed of the impact of progressive degradation and the possible ultimate failure of the structure itself in severe environmental conditions.

- As noted above, the design *S–N* curves have been developed over many years based on an increasing amount of test data. For ageing structures, the longer life part of these curves is particularly relevant. However, due to the need to test fatigue components in seawater at wave frequencies (i.e. 0.2 cycles per second), tests are very long and the data available for lives greater than 10^6–10^7 cycles are very limited, making this part of the *S–N* curve uncertain. It is assumed that for longer lives the *S–N* curve applicable to components in seawater with adequate cathodic protection reverts to the air curve. This, however, remains uncertain.

- The testing of components from decommissioned structures provides an opportunity to understand the actual performance of these components in the relevant environment and in the structure itself. A few tests have been undertaken to date but there is a considerable scope for further testing as more structures are decommissioned. Such tests should be used to verify the adequacy of current fatigue and strength assessment methodologies and for their further development, including the use of partial safety factors.

- There is some evidence that high strength steels (e.g. strengths >500 MPa) have inferior fatigue performance compared with the more conventional medium strength steels. This particularly applies when the cathodic protection levels are more negative than about −850 mV which can occur in practice, due to the difficulty of managing levels to be within the recommended range of −750 to −850 mV. However, the test data are limited. Hence, the assessment of ageing structures containing higher strength steels needs careful attention.

- The further development of continuous monitoring technology and autonomous inspection vehicles to provide improved and additional information on the condition of an ageing structure.

In addition, probabilistic methods provide a particularly useful approach for life extension assessment and managing uncertainties due to ageing but there are issues that need further research in order for such methods to become accepted as a basis for integrity evaluation:

- Probabilistic methods need to be improved, requiring particular focus on observable and measurable parameters (e.g. crack size) rather than abstract statistical quantities. In addition, the methods of updating in a probabilistic analysis of a structure should be based on actual data from the specific structure.

- Probabilistic inspection planning has for structures in their design life proved to be an acceptable method. However, in many cases probabilistic methods indicate increasingly longer inspection intervals for ageing structures while the likelihood of cracks is expected to increase. This is contradictory to the expectation of reduced inspection intervals needed for managing ageing structures. Further work is needed to improve the methodology on this topic.

- Further understanding of the management of structural integrity of jacket structures using predictions of fatigue life and redundancy is required, particularly taking into account the current practice of using the findings from flooded member detection (FMD) and general visual inspection (GVI), which only detect gross damage, and the expectation of increased degradation in ageing structures.

6.3 Final Thoughts

In the offshore industry there is already a large number of jackets, semi-submersibles, jack-ups, concrete structures and ship-shaped structures. Many of these structures are now ageing yet continue to make an important contribution to the production of oil and gas from existing fields. A key factor in managing life extension is balancing its cost with the associated commercial benefit of continued production.

Premature cessation of production from some structures would reduce total production capability unless new structures are commissioned. However, this would be unnecessarily costly if the existing structures are capable of extended life. In addition, this could be considered to be a misuse of construction materials and natural resources and hence the life extension of existing structures could reduce the environmental impact. Furthermore, the management of safe life extension is increasingly important and needs to be accepted by both operators and regulators.

Taking into consideration what we have discussed in this book about ageing structures, it would be useful to recommend two key factors that a designer of a new structure should take into account that would make a structure more suitable for life extension in the future. These are:

- Incorporation of sufficient redundancy and ductility to provide continuing integrity when degradation and damage begin to accumulate due to ageing.
- Design of all safety critical parts of the installation to be inspectable and repairable.

The continued safety of ageing offshore structures is an increasingly important issue for oil and gas companies and their service providers worldwide. It is recognised that life extension is a relatively new concept that highlights the issues of ageing and the methods for its justification are developing. Ultimately, the safety of personnel working on these ageing structures and any potential environmental damage resulting from an incident on an ageing structure are paramount.

If this book contributes to the understanding of the integrity management of offshore structures to avoid such consequences that may arise, then it will have achieved its objective in ensuring safe ageing offshore structures through their life extension phase.

Appendix A

Types of Structures

The types of platforms primarily used in the offshore industry are fixed and floating platforms. In addition, some examples of bottom supported exist, but these are not discussed in this book. Examples of offshore structure are shown in Figure A.1.

A.1 Fixed Platforms

An overview of fixed platforms is given here. These normally include:

- Steel jacket structures (mostly pile supported or suction anchor supported).
- Concrete gravity based structures.
- Jack-ups.

Jacket structures (Figure A.1b) are built on steel legs and piled to the seabed, supporting a deck with space for drilling rigs, production facilities, and crew quarters. Steel jackets are usually made of tubular steel members. A typical steel platform is shown in the figure below. Steel structures are subject to ageing processes such as fatigue and corrosion and hence life extension is a key issue with regard to the structure itself.

Concrete gravity based structures (Figure A.1c) typically consist of a base of concrete cells for oil storage and typically one to four shafts (also called caissons) that rise to above the sea surface for support of the topside.

Jack-ups (Figure A.1a) are self-elevating units with a buoyant hull and several legs which when on location can be lowered to the seabed and raise the deck above the level of the sea thus creating a more stable facility for drilling and/or production. Jack-up rigs have been primarily used for exploratory drilling but there are a few instances where they have also been used for production.

A.2 Floating Structures

Floating platforms are in general dependent on their watertight integrity and station keeping in addition to their structural integrity to function. The most used types of floating structures used in the offshore industry includes:

- Semi-submersible platforms (mostly in steel, but one in concrete exists)
- Tension leg platforms (TLPs) (mostly in steel, but one in concrete exists)

Ageing and Life Extension of Offshore Structures: The Challenge of Managing Structural Integrity, First Edition. Gerhard Ersdal, John V. Sharp, and Alexander Stacey. © 2019 John Wiley & Sons Ltd. Published 2019 by John Wiley & Sons Ltd.

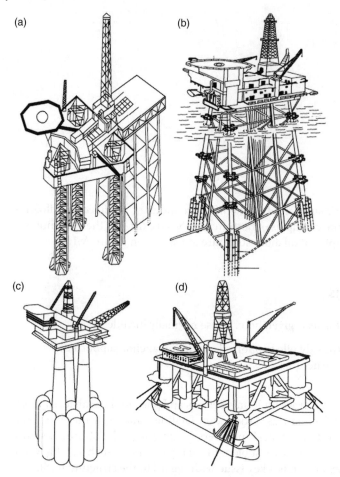

Figure A.1 Examples of types of offshore structures: (a) is a jack-up placed alongside a jacket structure; (b) is a jacket structure; (c) is a concrete gravity-based structure; and (d) is a semi-submersible unit. Source: Sundar (2015). Reproduced with permission of John Wiley & Sons.

- Ship-shaped platforms and barge shaped platforms (mostly in steel, but a few in concrete have been made)
- Spar platforms

Semi-submersible units (Figure A.1d) have hulls, together with columns and pontoons, with sufficient buoyancy to enable the structure to float. Semi-submersible platforms change draft by means of ballasting and de-ballasting (changing the water level in seawater tanks). They are normally anchored to the seabed by mooring systems, usually a combination of chain or wire rope. However, semi-submersibles also can be kept on station by dynamic positioning (DP). This is commonly used for semi-submersibles as drilling units and flotels in deep water. The hull supports a deck on which various facilities for drilling and/or production can be installed.

A TLP is similar to the semi-submersible platform but is vertically moored by tension legs which decreases the vertical motion significantly. A Spar platform is also to some extent similar to a semi-submersible, but consists of a single moored large-diameter vertical cylinder that is supporting the topside. The cylinder is ballasted in the lower part (often by solid and heavy material) to provide stability.

A ship-shaped structure, often called a floating storage unit (FSU), floating storage and offloading unit (FSO), and floating production, storage, and offloading unit (FPSO), is a floating vessel similar in shape to a conventional ship used by the offshore oil and gas industry for the production and processing of hydrocarbons. It is usually moored to the seabed by chains or wire ropes, although it can also be held on station using a dynamic positioning system.

Reference

Sundar, V. (2015). *Ocean Wave Mechanics: Application in Marine Structures*. Wiley.

Appendix B

Inspection Methods

This appendix summarises the basic information available on the inspection and monitoring techniques.

B.1 General Visual Inspection

General visual inspection (GVI) is a commonly applied inspection method used to detect large scale anomalies and to confirm the general condition and configuration of the structure. Underwater GVI is normally performed using a remote operated vehicle (ROV) without removal of marine growth. A benefit of the method is that it is possible to inspect large parts of the structure in a short time period.

B.2 Close Visual Inspection

Close visual inspection (CVI) is used to provide a more detailed examination of structural components or to examine suspected anomalies. Underwater CVI is normally undertaken using an ROV operated by a qualified inspector and usually requires removal of marine growth. CVI above water and internally inside dry compartments in floating structures should be performed by a qualified inspector. Dye penetration can be used to enhance the presence of cracks under dry conditions.

A benefit of the method is that a certain level of detailed inspection can be achieved in a reasonable time and for a reasonable cost.

B.3 Flooded Member Detection

Flooded member detection (FMD) can be used to detect the presence of water in initially air-filled submerged hollow members, indicating through-thickness cracks, using either a radioactive source or by ultrasonic methods. FMD is only used underwater and performed by an ROV. A benefit of the method is that it is possible to inspect large parts of the structure in a short time period.

Even if there is a through-thickness crack, flooding does not necessarily occur throughout the member particular in vertical and inclined members. This is especially a possible problem for members in compression.

Ageing and Life Extension of Offshore Structures: The Challenge of Managing Structural Integrity,
First Edition. Gerhard Ersdal, John V. Sharp, and Alexander Stacey.
© 2019 John Wiley & Sons Ltd. Published 2019 by John Wiley & Sons Ltd.

B.4 Ultrasonic Testing

Ultrasonic testing (UT) is primarily used for thickness measurements and is particularly used for monitoring components in floating structures for corrosion. Ultrasonic testing can be used underwater by an ROV, but more usually is used internally in hull and compartments, and also for topside inspection.

B.5 Eddy Current Inspection

Eddy current inspection (ECI) involves the use of magnetic coils to induce eddy currents in steel members. The presence of surface breaking cracks distorts the current field which can be detected by suitable probes. ECI can be used both in dry conditions and underwater. Very careful surface preparation is required for the method to be effective. ECI is capable of detecting quite small cracks, and according to DNVGL-RP-C210 there is a 90% probability of detecting a crack with a depth of 12 mm underwater and 3 mm in dry conditions.

ECI underwater is usually performed by divers, which has cost and safety implications.

B.6 Magnetic Particle Inspection

Magnetic particle inspection (MPI) involves the use of induced magnetic fields in a steel component and magnetic powder is used to identify the presence of surface breaking cracks. Very careful surface preparation is required. MPI is also capable of detecting quite small cracks, and probability of detecting a crack according to DNVGL-RP-C210 is similar to that of EC inspection.

MPI underwater is also usually undertaken by divers.

MPI requires removal of paint, and it is often difficult to reinstate the original corrosion protection. This may result in future corrosion attack, especially relevant for semi-submersible and ship-shaped structures.

B.7 Alternating Current Potential Drop

The alternating current potential drop (ACPD) technique is where an electric current is passed between two probes placed on the component. The presence of a surface breaking defect between the probes increases the electrical resistance locally. This change is used to detect and size defects.

B.8 Alternating Current Field Measurement

Alternating current field measurement (ACFM) is an electromagnetic technique used for the detection and sizing of surface breaking cracks in steel components. The method involves using a probe that introduces an electric current locally into the component and measures the associated electromagnetic field. A defect will disturb the associated field. The ends of a defect can be identified to provide information on defect location and length. ACFM inspection can be performed through paint and coatings, and therefore it is considered to imply less cost and time compared with several other non-destructive testing (NDT) techniques.

B.9   Acoustic Emission Monitoring

The principle of acoustic emission (AE) techniques is to use an arrangement of sensors to detect characteristic sound patterns that might signal the presence of structural anomalies locally in the structure. Acoustic emission has numerous applications, although the most relevant to offshore platforms is structural monitoring.

Acoustic emission can provide real time information on fatigue crack growth, and is primarily used to monitor known cracks. It is also claimed to be useful in detecting fatigue cracks at an early stage before conventional NDT methods would detect them. However, this would require a prearrangement of sensors in the relevant area.

B.10   Leak Detection

This monitoring system can involve a sensor to detect water and then raise an alarm via an audible or visual unit. Leak detection is particularly important for managing water tight integrity in floating structures. Typically, regulators and classification societies require an approved leak detection system for floating structures. This is particularly important when the fatigue utilisation index (FUI) exceeds 1.0.

B.11   Air Gap Monitoring

Monitoring the air gap establishes whether the platform foundations are suffering from subsidence, and also establishes the size of the air gap based on a prescribed maximum crest height. Modern methods use either Global Positioning System (GPS) or downward looking radar to monitor the air gap.

B.12   Strain Monitoring

Strain monitoring is carried out to measure the stress or loading regime in part of a structure. There are a number of strain measurement techniques available including conventional strain gauging, fibre optics, and stress probes.

Strain monitoring is primarily used in new types of structures to verify the design calculations. Strain monitoring can also be used to verify stresses in complex components in a life extension assessment. However, installation of strain monitoring in an existing structure is rather cumbersome and expensive.

B.13   Structural Monitoring

Structural monitoring is instrumented observation of a structure using sensors that continuously or periodically measure structural behaviour (such as dynamic frequency, strains, leakage, position and ultrasound from the structure) to identify changes to material or geometric properties and boundary conditions of the structure. Structural monitoring is described in more detail in Section 5.2.8.

Appendix C

Calculation Examples

C.1 Example of Closed Form Fatigue Calculation

Consider the case of a structure which has been in service for 20 years, and where the owner wants to extend the life for another 5–10 years. However, corrosion has been discovered on one of the fatigue prone members, and it has to be assumed that free corrosion is taking place and that the cathodic protection (CP) system is ineffective in this area.

If the stress histogram with stress blocks can be fitted to a Weibull distribution the Miner's sum can be shown to be (DNVGL RP-C203):

$$D = \int_{\Delta\sigma=0}^{\infty} \frac{n \cdot f(\Delta\sigma) \cdot d\sigma}{A_2/\Delta\sigma^m}$$

where $f(\Delta\sigma)$ is the frequency function fitted to the histogram and n is the total number of applied loading cycles. The Weibull model frequency function reads:

$$f(\Delta\sigma) = \frac{h}{q} \cdot \left(\frac{\Delta\sigma}{q}\right)^{h-1} \cdot \exp\left[-\Delta\sigma/q\right]^h$$

where h is the shape parameter and q is the scale parameter in the distribution. The Weibull distribution can be obtained both by cycle counting and by the energy-spectrum approach. The integral can then be determined by using the well-known Gamma function. For a single slope S–N curve the damage (D) can be calculated as:

$$D = \frac{n}{A_2} \cdot q^m \cdot \Gamma(1 + m/h)$$

where $\Gamma(.)$ denotes the Gamma function. This function can be found in standard tables and in mathematical computational programs.

In the case of a bilinear S–N curve the damage ratio will read (DNV-RP-C203):

$$D = \frac{n \cdot q^{m_1}}{\bar{a}_1}\Gamma\left(1 + \frac{m_1}{h}; \left(\frac{S_1}{q}\right)^h\right) + \frac{n \cdot q^{m_2}}{\bar{a}_2}\gamma\left(1 + \frac{m_2}{h}; \left(\frac{S_1}{q}\right)^h\right)$$

where $\bar{a}_1$ and m_1 are the fatigue parameters for the upper S–N line segment, $\bar{a}_2$ and m_2 are the fatigue parameters for the lower S–N line segment, S_1 is the stress level at

Ageing and Life Extension of Offshore Structures: The Challenge of Managing Structural Integrity,
First Edition. Gerhard Ersdal, John V. Sharp, and Alexander Stacey.
© 2019 John Wiley & Sons Ltd. Published 2019 by John Wiley & Sons Ltd.

the change in slope of S–N curves (point of discontinuity), n is the number of stress cycles, $\Gamma(x, y)$ is the complementary Gamma function, and $\gamma(x, y)$ is the incomplete Gamma function.

Assume that the detail can be calculated in accordance with the S–N curve E, and the number of cycles in 20 years is assumed to be $1*10^8$. the Weibull shape factor is further assumed to be $h = 0.9$ and the Weibull scale parameter is assumed to be $q = 9$ (i.e. the maximum stress range in the data set is $S_{max} = 230$ MPa).

In this example the S–N curve E based on DNVGL RP-C203 gives $\log(a_1) = 11.61$, $\log(a_2) = 15.35$ and $S_1 = 74.13$ MPa when the structure is protected by CP. The calculated fatigue damage in 20 years is then $D = 0.7$ (annual damage of $D = 0.035$) indicating a few years potential life remaining for this specific detail. However, the observed corrosion indicates that the free corrosion S–N curve needs to be used for the last three years.

The S–N E curve with free corrosion in DNVGL RP-C203 with $\log(a) = 11.533$ gives a yearly damage of $D = 0.1$. Hence, at the time of the assessment $D \approx 0.9$ indicating a possible life extension for this detail of only one year.

An important observation from this example is the increase in yearly fatigue damage from the lack of corrosion protection (from $D = 0.035$ to $D = 0.1$).

C.2   Example of Application of Fracture Mechanics to Life Extension

Consider the case of a semi-submersible which had been in service for 25 years and has reached the end of its design fatigue life. The operator has decided to perform a detailed inspection and assessment of the structure to determine the viability of extending the operating life of the structure. The structure was dry-docked and comprehensive inspection, in accordance with the requirements of the classification society, revealed that a number of defects were located at the weld toes of a number of joints.

Weld grinding and further inspection by magnetic particle inspection (MPI) and ultrasonic testing (UT) were performed to verify that the defects had been removed. Guidance on grinding is given in DNVGL-RP-C203.

Fracture mechanics assessments were performed on both the ground joints and those with the lowest fatigue lives to determine whether the semi-submersible would be likely to experience significant cracking resulting in structural failure in the five-year interval to the next major inspection in dry dock.

The fracture mechanics assessment of each joint entails the following steps:

1. Initial information
   - Identification of the geometrical parameters, including the member thickness.
   - Determination of the hot spot stress fatigue loading at the most highly stressed joint locations from a finite element analysis of the structure using the latest metocean data.
   - Determination of the static loading corresponding to the extreme storm (100-year) event.

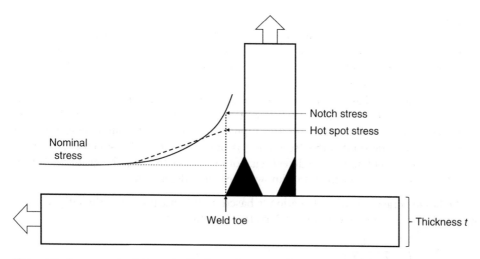

Figure C.1 Components of stress distribution acting on crack.

- Determination of the fracture toughness of the material, entailing crack tip opening displacement (CTOD) testing of both the heat affected zone (HAZ) and the weld metal to obtain the minimum value.
- Establishment of the minimum detectable defect size for the non-destructive testing methods used.

2. Evaluation of local weld stress concentration factors

   The total crack-opening stress needs to be used in the fracture mechanics assessment. This is determined from the superposition of the hot spot stress and the stress distribution due to the stress concentration at the weld toe, i.e. the notch stress, as shown in Figure C.1.

   For welded joints, a stress concentration factor (SCF) of three at the weld toe is commonly used. This SCF decays sharply beyond the vicinity of the weld toe but has a significant influence on small defects within its range and contributes to the initiation and propagation of defects at the weld toe by enhancing the fatigue stresses in the joint at the weld toe. Grinding not only removes microscopic defects which can develop into fatigue cracks but also mitigates crack initiation and propagation as weld profiling reduces the SCF.

3. Stress intensity factor, K, solution

   The stress intensity factor, K, solution used to evaluate the ΔK range in the fatigue crack growth assessment and the maximum K for the fracture assessment is dependent on the geometry and a range of solutions are available. K solutions for a specific geometry can be derived using finite element analysis. However, this is a very resource-intensive process and for complex geometries it is usual to use solutions derived from simple geometries, such as plates and cylinders, with suitable modifications to represent the weld SCF.

4. Fatigue crack growth and fracture analysis

The fatigue crack growth analysis predicts the time for a known or assumed defect to propagate to a specified or critical size or it can be used to determine the crack growth for a particular period of time, e.g. the inspection interval, and whether failure would be expected during this time.

The fracture mechanics analysis is a complex process requiring the use of a computer program into which a defect assessment method, e.g. BS 7910, is coded. It requires careful selection of the input parameters. The initial defect size is particularly important if the assessment is based on a postulated defect size, as in the case of a weld in which a defect has not been detected (e.g. the ground welds in this case study), rather than on a known/measured defect size.

The fatigue crack growth calculation is based on the integration of the fatigue crack growth law which, in the case of the Paris law, is as follows:

$$N = \int_{a_i}^{a_f} \frac{da}{C \cdot (\Delta K)^m}$$

The numerical calculation entails the summation of a large number, n, of small increments of crack growth, Δa, i.e.

$$N = \int_{a_i}^{a_f} \frac{da}{C \cdot (\Delta K)^m} = \sum_i^n \Delta N_i = \sum_i^n \frac{\Delta a_i}{C \cdot [(\Delta K_i)]^m}$$

where

$$\Delta K_i = Y_i(\Delta\sigma)\sqrt{(\pi a_i)}$$

and $\Delta\sigma$ represents the fatigue loading spectrum.

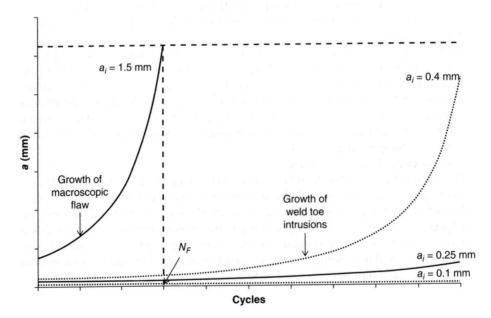

Figure C.2 Influence of initial crack size on remaining life predictions.

The number of increments needs to be sufficiently large so that convergence of the computed result is achieved. At each increment of crack growth, the program performs a fracture check based on the failure assessment diagram (FAD). This requires both the application of a suitable CTOD value for the K_r calculation and the selection of the appropriate plastic collapse solution for the L_r calculation.

Figure C.2 shows the effect of the assumed initial crack size on the growth of the crack. It is clear that small differences in the assumed initial crack size (ranging from 0.1 to 1.5 mm) can have a very significant impact on the predicted remaining life and this demonstrates the importance of the selection of the initial crack size.

Index

Ageing and Life Extension of Offshore Structures: The Challenge of Managing Structural Integrity,
First Edition. Gerhard Ersdal, John V. Sharp, and Alexander Stacey.
© 2019 John Wiley & Sons Ltd. Published 2019 by John Wiley & Sons Ltd.